Advances in Aquatic Ecology

— Volume 2 —

Advances in Aquatic Ecology

— Volume 2 —

Editor
Dr. Vishwas B. Sakhare
Post Graduate Department of Zoology
Yogeshwari Mahavidyalaya,
Ambajogai – 431 517
Maharashtra

2008
DAYA PUBLISHING HOUSE
Delhi - 110 035

ISBN10 81-7035-559-1
ISBN13 81-7035-559-5

Published by : **Daya Publishing House**
1123/74, Deva Ram Park
Tri Nagar, Delhi - 110 035
Phone: 27383999
Fax: (011) 23260116
E-mail: dayabooks@vsnl.com
website: www.dayabooks.com

Showroom : 4760-61/23, Ansari Road, Darya Ganj,
New Delhi - 110 002
Phone: 23245578, 23244987

Laser Typesetting : **Classic Computer Services**
Delhi - 110 035

Printed at : **Chawla Offset Printers**
Delhi - 110 052

PRINTED IN INDIA

Dedicated to
My Father

Shri Balasaheb Pandharinath Sakhare

Managing Director
Sagar Co-operative Sugar Factory Ltd.
District Jalna (M.S.)

Preface

The second volume of *Advances in Aquatic Ecology* is compendium of recent original research in the field of aquatic ecology. It is an assemblage of up-to-date information of rapid advances and developments taking place in the field of aquatic ecology.The book is a unique compilation of 20 chapters which discusses exhaustic studies on planktonology, physico-chemical environment, fisheries, toxicology, physiology etc.with its application oriented and interdisciplinary approach, the book would be immensely useful to students, teachers, researchers, scientists, policy makers, environmental lawyers and others interested in aquatic ecology. The chapters in the book have been contributed by eminent scientists/academicians active in the areas of aquatic ecology.

My special thanks and appreciation go to the scientists whose contributions have enriched this volume. I wish to express my sincere gratitude to Dr.Sureshji Khursale, President, Yogeshwari Education Society, Ambajogai who has been a source of constant inspiration.I am especially thankful to Dr.D.H.Thorat, Principal, Yogeshwari Mahavidyalaya, Ambajogai for his encouragement. I owe my special thanks to Dr.P.K.Joshi of Dnyanopasak College, Parbhani, Dr. Indranil Ghosh of West Bengal University of Animal and Fishery Sciences, Kolkata, Dr.Prakash Shingare of Khar Land Research Station of B.S.Konkan Krishi University, Panvel, Dr.Mamta Rawat of J.N.V.University, Jodhpur, Dr.H.M.Ashashree of Kuvempu University, Shimoga, Dr.M.Rajkumar and Surajit Das of Annamalai University, Dr.Sachin Satam of Marine Products Export Development Authority (MPEDA), Mumbai, Dr.R.J.Chavan of Arts, Commerce and Science College, Kille Dharur (Maharashtra) and Dr.K.Pazhanisamy of Government Arts College, Ariyalur (Tamil Nadu).

I wish to thank my wife Surekha for her endurance during the compilation work of this volume. She has helped me for constantly data feeding, processing and reprocessing on computer. I want to thank my father Shri Balasaheb Sakhare, brother Avinash, sisters Meenal and Preeti for their help in many ways. I thank my publisher Shri Anil Mittal of Daya Publishing House, Delhi for taking pains in bringing out the book.

Finally, I will always remain a debtor to all my well-wishers for their blessings, without which this volume would not have come into existence.

Dr. V.B. Sakhare

Contents

List of Contributors

Ashashree, H.M.
Department of Zoology, P.G. Studies and Research Dept. of Applied Zoology, Kuvempu University, Shankarghatta, Shimoga, Karnataka
E-mail: hm_ashree@yahoo.co.in

Babare, M.G.
PG and Research Department of Zoology, Arts, Science and Commerce College, Naldurg – 413602, Maharashtra

Chavan, S.P.
P.G. and Research, Department of Zoology and Fisheries, Dnyanopasak College, Parbhani – 431401, Maharashtra

Das, Surajit
Centre of Advanced Study in Marine Biology, Annamalai University, Parangipettai – 608 502, Tamil Nadu
E-mail: surajit@myself.com

Dhane, S.S.
Khar Land Research Station, Dr.B.S.Konkan Krishi Vidyapeeth, Bander Road, Panvel – 410 206, Maharashtra

Dookia, Sumit
Ecology and Rural Development Society, C–123, Krishna Nagar, Jodhpur – 342005, Rajasthan

Girkar, Milind M.
C/o, Vijay Rane, 7/20, Tardeo Police Camp, Tardeo, Mumbai – 400 034

Hosmani, S.P.
Department of Studies in Botany, Manasgangotri, University of Mysore, Mysore, Karnataka

Indumathy, P.
Department of Zoology, Queen Mary's College, Chennai – 600 004, Tamil Nadu

Jakher, G.R.
Department of Zoology, J.N.V.University, Jodhpur – 342 005, Rajasthan

Jayanthi, J.
Department of Advanced Zoology and Biotechnology, Guru Nanak College, Chennai – 600 042, Tamil Nadu

Kadam, M.S.
P.G. and Research Department of Zoology, Yeshwant College, Nanded – 431602, Maharashtra

Kanna, L
Thiruvalluvar University, Fort Campus, Vellor – 632 004, Tamil Nadu

Khan, S. Ajmal
Centre of Advanced Study in Marine Biology, Annamalai University, Parangipettai – 608 502, Tamil Nadu

Lakshmi Devi, P.
Department of Zoology, Queen Mary's College, Chennai – 600 004, Tamil Nadu

Lyla, P.S.
Centre of Advanced Study in Marine Biology, Annamalai University, Parangipettai – 608 502, Tamil Nadu

Mehta, V.B.
Ex–Khar Land Scientist, Khar Land Research Station, Dr. B.S. Konkan Krishi Vidyapeeth, Bander Road, Panvel – 410 206, Maharashtra

Meshram, S.J.
Khar Land Research Station, Dr. B.S. Konkan Krishi Vidyapeeth, Bander Road, Panvel – 410 206, Maharashtra

Mote, M.V.
PG and Research Department of Zoology, Arts, Science, and Commerce College, Naldurg – 413 602, Maharashtra

Pazhanisamy, K.
Lecturer in Zoology, Government Arts College, Ariyalur – 621 713, Tamil Nadu
E–mail: pazhani_zooo@yahoo.com

Perumal, P.
Centre of Advanced Study in Marine Biology, Annamalai University, Parangipettai – 608 502, Tamil Nadu
E–mail: perumal_dr@yahoo.co.in

Ragunathan, M.G.
Department of Advanced Zoology and Biotechnology, Guru Nanak College, Chennai – 600 042

Rajalakshmi, S.
KMCPG Studies, Lawspet, Pondicherry

Rajkumar, J.
Centre of Advanced Study in Marine Biology, Annamalai University, Parangipettai – 608 502
Tamil Nadu

Rajkumar, M.
Centre of Advanced Study in Marine Biology, Annamalai University, Parangipettai – 608 502
Tamil Nadu
E–mail: arunachalashivamdr@yahoo.com

Rawat, Mamta
Department of Zoology, J.N.V. University, Jodhpur – 342 005, Rajasthan
E–mail: rawatscorner@yahoo.co.in

Reddy, K.R.
PG and Research Department of Zoology, Arts, Science and Commerce College, Naldurg – 413 602, Maharashtra

Sahu, Maloy Kumar
Centre of Advanced Study in Marine Biology, Annamalai University, Parangipettai – 608 502, Tamil Nadu

Sangeetha, M.
Department of Zoology, Queen Mary's College, Chennai – 600 004, Tamil Nadu

Satam, Sachin B.
State Coordinator (OFD), Marine Products Export Development Authority (MPEDA), RO, Mumabi – 400 021, Maharashtra
E–mail: sachin_satam@rediffmail.com

Shingare, Prakash
Associate Professor, Kharland Research Station, Panvel, Distt. Raigad, Maharashtra
E–mail: prakashshingare@yahoo.com; prakashshingare@gmail.com

Shirgur, G.A.
Ex–Research Officer, Taraporwala Marine Biological Research Station, Mumbai, Maharashtra

Sivakumar, K.
Centre of Advanced Study in Marine Biology, Annamalai University, Parangipettai – 608 502, Tamil Nadu

Sree Latha, K.
Department of Zoology, Dr. S.R.K. Government Arts College, Yanam – 533 464, (U.T. of Pondicherry)

Srikantaswamy, S.
Department of studies in Environmental Science, Manasgangotri, University of Mysore, Mysore, Karnataka

Sudeep, B.M.
Department of studies in Botany, Manasgangotri, University of Mysore, Mysore, Karnataka

Swamy, H.M. Renuka
Department of Zoology, P.G. Studies and Research Dept. of Applied Zoology, Kuvempu University, Shankarghatta, Shimoga, Karnataka

Thangaradjou, T.
Centre of Advanced Study in Marine Biology, Annamalai University, Parangipettai – 608 502, Tamil Nadu

Thavasi, R.
Centre of Advanced Study in Marine Biology, Annamalai University, Parangipettai – 608 502, Tamil Nadu

Triles, J.P.
Equipe Adaptation Ecophysiologique et Ontogenes, UMR 5171 (UMZ–IFREMER–CNRS) "Genome, Populations, Interactions, Adaptations", Universite' Montpellier II, CPO92, Nace Eugene Bataillon, 34095, Montpellier Gedex 5, France

Vasagam, K.P. Kumaraguru
Centre of Advanced Study in Marine Biology, Annamalai University, Parangipettai – 608 502, Tamil Nadu

Venkateshwalu, M.
Department of Zoology, P.G. Studies and Research Dept. of Applied Zoology, Kuvempu University, Shankarghatta, Shimoga, Karnataka

Chapter 1

Biochemical Composition of Wild Copepods, *Acartia erythraea* Giesbrecht and *Oithona brevicornis* Giesbrecht from Coleroon Coastal Waters, Southeast Coast of India

M. Rajkumar, K.P. Kumaraguru Vasagam and P. Perumal

Introduction

Information on biochemical composition of copepod is of great importance in understanding the nutritive value and energy pathways in the pelagic food web. The marine copepods are considered to be "nutritionally superior" as food for marine fish and prawn larvae as they are a valuable source of proteins, carbohydrate, lipids, amino acids, fatty acids and enzymes (amylase, protease, exonuclease and esterase), all of which play an important role in larval digestion and enhance metamorphosis of larvae (Fluchter and Rembold, 1986; Munilla-Moran *et al.*, 1990; Millamena *et al.*, 1990 and Støttrup, 2000). Most of the fish and crustacean species mainly depend on zooplankton (copepods) throughout their life stages and some species even feed exclusively on copepods during their entire life. In aquaculture, larval pigmentation, growth and survival of fish are normally satisfactory only when copepods are used as live-feed. Studies on protein, carbohydrate and lipid content in the various zooplankton groups imply that these substances may have an important role in the physiological functions, metabolism, nutritive value and reproductive and energetic aspects of the marine ecosystem (Bhat and Wagh, 1992; Sreepada *et al.*, 1992; Krishnakumari and Goswami, 1993; Maruthanayagam and Subramanian, 1999; Nageswara Rao and Krupanidhi, 2001 and Ashok Prabu *et al.*, 2005).

Copepods are the most abundant and probably the most ecologically significant animals of the first consumer level of the marine plankton. In aquaculture chilled and frozen zooplankters are often used and they float, making it easier for the fish to catch (Tucker, 1992). The live-feed organisms like copepods are called "living capsules of nutrition" for fish and prawn larvae (Thakur, 1998). The high content of amino acids (Dabrowski and Rusiecki, 1983), enzymes (Lauff and Hofer, 1984) and water (Lemm, 1983) in zooplankton are all positive qualities for starting feeding by fish larvae (Holm and Moller, 1984). Free amino acids are present in the frozen fluid that surrounds the zooplankton and these form a powerful attractant and appetite stimulant for fish larvae (Dabrowski and Rusiecki, 1983; Mearns, 1986 and Tucker, 1992). Further, the nutritional quality of copepods is important in determining the survival and growth of fish larvae (Szyper, 1989). In many marine fish and shrimp hatcheries, high mortality rates during the first feeding of larvae are linked to lack of nutritionally inadequate live-food organism. Copepods have been shown to be essential diet for solving these problems.

The limited information available on biochemical composition of wild zooplankton in India is related to the general quantitative aspects of protein, fat and carbohydrate (Madhupratap *et al.*, 1979; Rosamma Stephen *et al.*, 1979; Goswami *et al.*, 1981; Sumitra Vijayaraghavan *et al.*, 1982; Krishnakumari and Nair, 1988; Nandakumar *et al.*, 1988; Krishnakumari and Achuthankutty, 1989; Bhat and Wagh, 1992; Ingole and Parulekar, 1995; Goswami *et al.*, 2000 and Ashok Prabu *et al.*, 2005) as well as the amino acid content of cultured copepod *Oithona rigida* (Santhanam and Perumal, 2001 and Rajkumar and Kumaraguru Vasagam, 2006).

Marine copepods have paramount ecological significance as they form pelagic food links and they constitute the largest and reliable source of protein. With all essential amino acids, they play an important role in the growth and survival in fish and crustacean larvae when used as feed in aquaculture. Amino acids are served as an energy substrate in the developing eggs and larvae of fishes. Only very limited works are available on the amino acid composition of copepods (Yurkowski and Tabachek, 1979; Dabrawski and Rusiecki, 1983 and Santhanam and Perumal, 2001). Hence in the present study, we focused on the seasonal variations in the biochemical composition *viz.*, protein, lipid, carbohydrate, moisture, ash contents and amino acid profiles of two wild copepods the calanoid copepod *Acartia erythraea* and the cyclopoid copepod *Oithona brevicornis* collected from 4 locations Bay of Bengal, Coleroon mouth, Coleroon estuary and Vettar backwaters.

Materials and Methods

Study Area

The river Coleroon, a branch of the river Cauvery, originates from Brahmagiri of the Western ghats and after meandering for over a distance of 760 kms, forms a very fertile delta in Nagapattinam District in Tamil Nadu before it debouches into the Bay of Bengal at a place called Pazhayaru (Lat. 11° 21′ N; Long. 79° 50′ E) on the south east coast of India. Pazhayaru is one among the three major fishing harbours in Tamil Nadu. It contributes a good quantity of seafood for the state. The river Coleroon flows into the Bay of Bengal at about 10 kms south of Parangipettai. The average width of the Coleroon-estuary varied from 420 meters in mouth region to 100 meters in upstream and the average depth near the mouth was about 7 meters and 1 or 2 meters in the tidal and freshwater zones. The Coleroon estuary has her estuarine connections with Pichavaram mangroves in the northern side and Buckingham canal and Vettar backwaters near the mouth on the southern side. The Buckingham canal and Vettar backwaters are the two small adjacent drainage channels of paddy fields. The mouth of the Coleroon river is always open and it shows semidiurnal tide and the tidal flushing extends up to a distance of about 15 kilometers upstream.

The present study was carried out from four different environments *viz*: (*i*) The sea (*ii*) Coleroon Mouth (*iii*) Estuary and (*iv*) The Vettar backwater for a period of one year (April 2004–March 2005). Station I was in 10 fathom line of Bay of Bengal and the Station II was in the mouth of the Coleroon river. This station is highly influenced by freshwater inflow during northeast monsoon (October-December). The mean salinity was 33 per cent. The bottom is characterized by sand. Station III was situated near the mouth region of the estuary. The bottom was characterized with mud deposits. Station IV was located at about 1.5 km from the Station II, which is highly influenced by freshwater through the adjacent Vettar backwater. The Vettar backwater has also got connection with two channels *viz.*, 'Chinna Vettar' and 'Semmangadu channel' (Figure 1.1).

Collection

Copepod (zooplankton) samples were collected from the study areas by horizontal-surface towing of a plankton net (0.35 cm mouth diameter), made up of bolting silk cloth (No 10: mesh size 158 μm) for 30 minutes. The copepod samples were immediately transported to laboratory and thoroughly rinsed to reduce the contamination from other zooplankters. From the samples, the copepods were identified

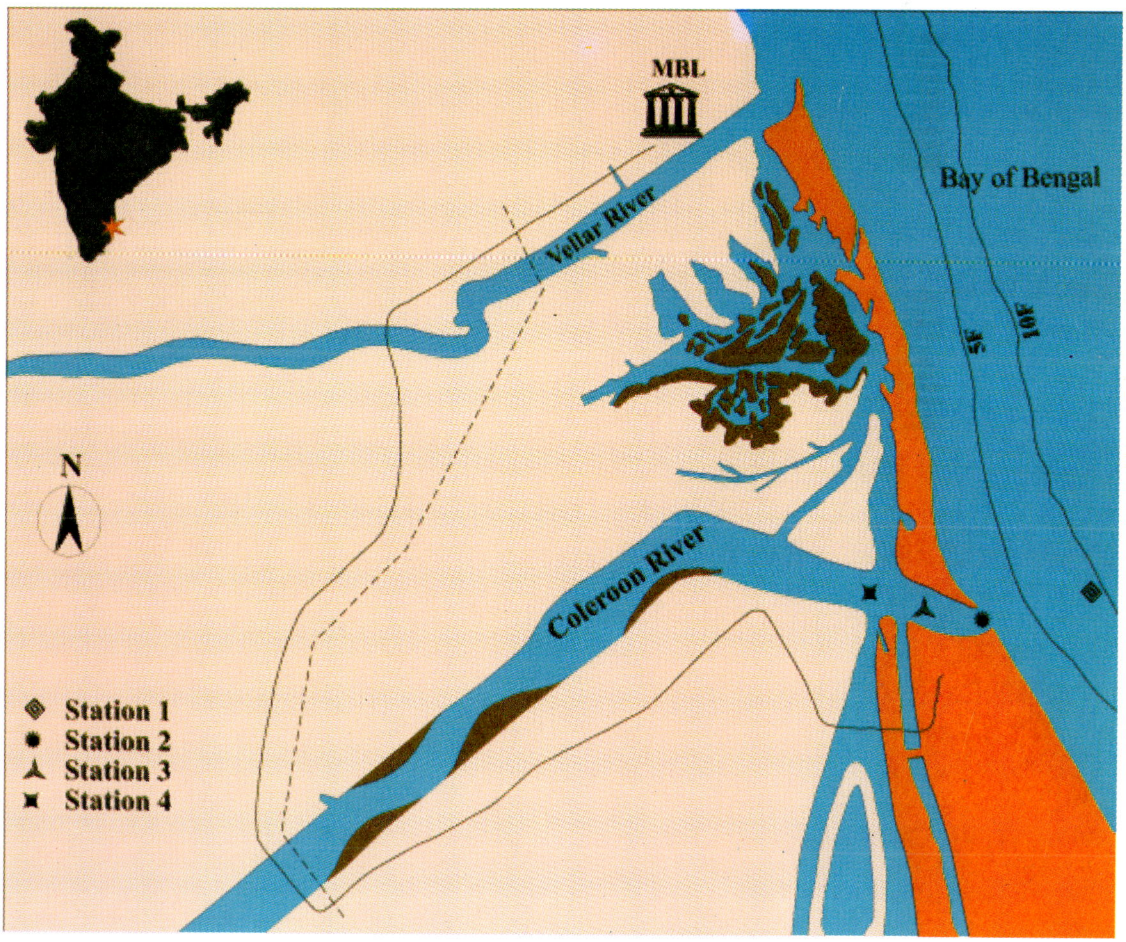

Figure 1.1: Map Showing the Study Area

under the microscope using the key of Kasturirangan (1963). The copepod samples were sorted out through the microscopical investigation and they were analyzed biochemically. After recording biomass (displacement volume), the samples were cleared of debris, washed with distilled water. Half of each sample was preserved for taxonomical studies and the rest was fresh dried (60°C) until constant weight was obtained. Triplicate analysis was made and the average value taken.

Isolation

After the collection, the zooplankton was screened to isolate the size fraction containing dominantly adult and later stage copepodites of *A. erythraea* and *O. brevicornis*. The zooplankton samples were first screened coarsely through a 500 µm mesh to remove the larvae of fish and prawn. Then they were rinsed for 2 hours in a zooplankton washer (Schipp *et al.*, 1999) fitted with a 190µm mesh screen used to remove smaller zooplankters such as rotifers, copepod nauplii and barnacle nauplii. The vigorous rinsing was made continuously for 2 hrs, after which most of the rotifer contamination was reduced and the rinsing was done periodically to reduce rotifer numbers. After rinsing, the remaining adult copepods were used for biochemical composition.

Biochemical Composition

Triplicate analysis was made and the average values were taken. Proximate composition were analysed for their protein, dry matter, ash content following (AOAC, 1995). Carbohydrate and lipid were analysed following Dubois *et al.* (1956) and Folch *et al.* (1956) respectively.

Amino Acid Estimation

The protein bound amino acids were separated from free amino acids by using 6 per cent TCA solution (Finn *et al.*, 1995). Amino acids were analysed after sealed tube hydrolysis with 6N HCL for 22 h at 110°C (Spackman *et al.*, 1958 and Finlayson, 1964). After hydrolysis, the acid was evaporated in vacuum oven and the sample was kept in a NaOH desiccator to remove traces of acid. The residue was brought into 1 ml of sample diluent (pH 2.20). Amino acids were analysed using the Shimadzu HPLC model LC-10A (Shimadzu corp., Japan). Separation of amino acid was performed in a column (Shimpack ISC-07/S 1504 Na) packed with a strongly acidic Na^+ type cation exchange resin (Styrene-divinyl benzene copolymer with sulfinic group) under gradient elusion. The amino acids were detected and quantified using a fluorescent detector (FLD-6A) after post column derivitization with O-pthaladehyde and 2-mercaptoethanol. Amino acid standard solution (Sigma-aldrich Inc., USA) for fluorescent detection was used as external standard. For every ten-sample injection, one standard run was carried out. The n mole units of amino acids were converted to µg units by multiplying with its molecular weight X 1000.

Results

Proximate Composition of *A. erythraea*

The concentration of protein was higher than the concentrations of lipid and carbohydrate. Moisture, protein, lipid, Carbohydrate and ash contents (per cent dry weight) of *A. erythraea* collected from the Coleroon coastal waters are given in Tables 1.1-1.4. At stn 1, the moisture, ranges (per cent) of protein, lipid, carbohydrate and ash were: 83.13-85.27; 70.52-74.43; 13.65-15.90; 6.02-7.35 and 3.21-4.22 respectively. At station 2, the ranges of percentage of moisture, protein, lipid, carbohydrate and ash were: 83.63-85.05; 68.99-73.76; 13.49-17.55; 4.59-6.30 and 3.64-4.20 respectively. At station 3, the moisture, ranges (per cent) of protein, lipid, carbohydrate and ash were: 83.98-85.17; 69.58-74.06; 14.26-17.53; 4.70-6.50 and 3.44-4.32 respectively. At station 4, the ranges of percentage of moisture,

Table 1.1: Biochemical Composition of the Copepod, *A. erythraea*, at Station 1 (Bay of Bengal)*

Parameter (%)	*April, 2004*	*May*	*June*	*July*	*Aug.*	*Sep.*
Moisture	85.2±0.37	85.14±0.37	85.01±0.18	85.20±0.18	85.08±0.45	84.78±0.36
Protein	73.07±1.37	73.05±1.13	74.43±1.27	74.40±0.82	74.15±0.89	73.86±1.01
Lipid	15.46±1.38	15.72±1.25	14.60±1.25	14.99±1.84	15.25±1.88	14.66±0.90
Carbohydrate	6.31±1.79	7.08±1.17	6.99±0.02	6.42±1.29	6.40±1.22	7.35±0.23
Ash	3.63±0.11	3.33±0.11	3.60±0.18	3.68±0.20	3.21±0.46	3.49±0.22
Parameter (%)	*Oct., 2004*	*Nov.*	*Dec.*	*Jan., 2005*	*Feb.*	*Mar.*
Moisture	83.77±0.85	83.13±1.65	85.27±0.27	84.71±0.30	84.60±0.74	84.37±0.30
Protein	71.88±1.44	70.52±1.64	72.49±0.19	72.71±1.62	70.91±3.33	73.44±1.09
Lipid	13.65±1.80	13.76±1.08	14.19±0.66	15.90±0.61	15.71±0.84	16.78±1.26
Carbohydrate	6.02±1.61	6.72±1.23	6.56±1.37	6.77±1.60	6.88±1.07	7.29±0.14
Ash	4.12±0.15	3.99±0.14	4.22±0.52	3.89±0.29	3.98±0.54	3.65±0.22

*: Means of triplicate samples ± standard deviation.

Table 1.2: Biochemical Composition of the Copepod, *A. erythraea*, at Station 2 (Mouth of the Coleroon River)*

Parameter (%)	*April, 2004*	*May*	*June*	*July*	*Aug.*	*Sep.*
Moisture	84.92±0.36	84.63±0.24	85.01±0.44	84.66±0.45	84.70±0.56	84.59±0.08
Protein	72.56±0.13	72.28±0.53	72.69±0.50	73.73±0.55	72.98±0.23	73.76±1.01
Lipid	17.26±0.39	17.27±0.60	16.71±0.53	16.84±0.57	16.13±1.13	16.10±0.99
Carbohydrate	4.83±0.17	5.33±0.13	6.00±1.06	4.86±0.24	5.12±0.77	5.12±0.36
Ash	3.74±0.13	3.77±0.56	3.84±0.32	3.64±0.55	3.88±0.56	4.12±0.43
Parameter (%)	*Oct., 2004*	*Nov.*	*Dec.*	*Jan., 2005*	*Feb.*	*Mar.*
Moisture	83.69±1.83	84.73±1.57	84.55±0.16	83.63±0.46	84.19±0.18	85.05±0.74
Protein	68.99±1.22	70.70±0.98	69.51±2.77	72.57±1.08	71.26±2.45	72.86±1.96
Lipid	14.86±0.67	13.49±0.97	14.01±0.60	16.72±0.39	15.83±1.41	17.55±0.32
Carbohydrate	5.51±1.35	4.59±0.16	5.59±1.05	6.12±1.06	5.76±1.22	6.30±1.41
Ash	4.20±0.52	4.17±0.46	3.53±0.12	4.08±0.34	3.82±0.25	3.88±0.31

*: Means of triplicate samples ± standard deviation.

Table 1.3: Biochemical Composition of the Copepod, *A. erythraea,* at Station 3 (Coleroon Estuary)*

Parameter (%)	*April, 2004*	*May*	*June*	*July*	*Aug.*	*Sep.*
Moisture	85.17±0.35	84.58±0.16	84.87±0.31	85.00±0.51	84.69±0.56	84.49±0.09
Protein	72.48±0.10	72.37±0.38	72.90±0.42	73.70±0.57	73.07±0.21	73.32±1.13
Lipid	17.19±0.43	17.13±0.53	16.33±0.45	16.93±0.58	16.88±0.90	16.11±0.98
Carbohydrate	4.70±0.38	5.45±0.29	6.50±1.06	4.98±0.17	5.27±0.64	6.00±1.24
Ash	3.68±0.04	3.44±0.18	3.76±0.40	3.61±0.54	3.45±0.69	3.85±0.49
Parameter (%)	*Oct., 2004*	*Nov.*	*Dec.*	*Jan., 2005*	*Feb.*	*Mar.*
Moisture	83.98±1.80	84.60±1.50	84.75±0.27	84.09±0.46	84.12±0.23	85.11±0.65
Protein	69.58±1.29	69.99±1.52	71.12±2.39	73.20±1.59	72.41±2.69	74.06±1.24
Lipid	15.18±0.81	14.26±0.87	14.32±0.55	16.33±0.67	16.42±1.05	17.53±0.30
Carbohydrate	4.88±0.89	4.70±0.53	5.06±1.42	5.58±1.54	5.20±1.30	5.67±1.81
Ash	4.32±0.33	3.99±0.38	3.72±0.25	4.16±0.33	3.78±0.29	3.61±0.20

*: Means of triplicate samples ± standard deviation.

Table 1.4: Biochemical Composition of the Copepod, *A. erythraea*, at Station 4 (Vettar Backwater)*

Parameter (%)	*April, 2004*	*May*	*June*	*July*	*Aug.*	*Sep.*
Moisture	85.08±0.29	84.73±0.40	84.88±0.31	84.95±0.45	84.75±0.55	84.91±0.63
Protein	72.30±0.37	72.04±0.95	72.52±0.71	73.40±0.90	72.65±0.71	73.62±0.99
Lipid	16.62±1.26	16.40±1.32	15.55±1.66	15.80±1.91	15.85±1.52	15.55±1.62
Carbohydrate	4.70±0.38	5.22±0.19	5.42±1.72	4.80±0.31	5.06±0.84	5.05±0.47
Ash	3.67±0.04	3.45±0.18	3.87±0.31	3.53±0.54	3.73±0.50	3.77±0.59
Parameter (%)	*Oct., 2004*	*Nov.*	*Dec.*	*Jan., 2005*	*Feb.*	*Mar.*
Moisture	83.83±1.80	84.24±1.47	84.99±0.66	84.40±0.96	84.12±0.22	85.21±0.48
Protein	69.42±1.18	70.66±0.98	70.50±2.15	72.56±1.08	72.06±2.45	73.71±1.15
Lipid	14.08±1.58	13.78±0.67	14.32±0.55	6.63±0.37	16.17±1.11	17.35±0.25
Carbohydrate	4.57±1.14	4.22±0.60	5.68±1.09	6.15±1.04	5.54±1.01	5.79±2.10
Ash	4.39±0.22	4.06±0.38	3.67±0.19	3.97±0.43	3.89±0.19	3.76±0.13

*: Means of triplicate samples ± standard deviation.

Table 1.5: Seasonal Variations in Amino Acid Composition of *A. erythraea* Collected from Coleroon Coastal Waters

Amino Acid	*A. erythraea (Sea)*				*A. erythraea (Mouth)*				*A. erythraea (Estuary)*				*A. erythraea (Vettar Backwaters)*			
	M	*PM*	*S*	*PRM*	*M*	*PM*	*S*	*PRM*	*M*	*PM*	*S*	*PRM*	*M*	*PM*	*S*	*PRM*
Aspartic acid	8.481	7.017	7.410	6.595	7.482	5.676	8.551	7.637	7.212	8.443	6.614	6.009	6.981	7.417	6.673	6.065
Threonine	15.215	16.786	15.839	14.656	16.659	10.953	16.020	16.422	17.529	15.752	13.778	16.602	16.655	14.686	14.739	15.556
Serine	8.961	8.715	9.960	7.604	7.769	7.260	9.396	8.963	5.422	6.404	7.124	7.339	5.661	7.415	8.060	6.594
Glutamic acid	14.621	13.977	11.459	14.073	12.327	11.944	14.998	12.578	12.409	14.396	18.432	13.314	12.221	12.277	16.359	12.273
Glycine	3.794	4.793	5.496	4.293	3.790	3.386	4.236	5.112	5.508	3.433	4.604	3.862	4.494	4.393	3.596	4.793
Alanine	8.315	10.292	9.352	7.884	8.505	12.209	8.839	10.071	8.781	9.344	9.745	7.802	7.715	10.292	8.652	6.983
Cystine	Trace	Trace	Trace	2.010	Trace	Trace	Trace	Trace	Trace	Trace	Trace	Trace	Trace	Trace	Trace	Trace
Valine	6.881	7.384	7.913	7.406	7.582	10.264	7.434	7.863	8.222	8.426	7.004	7.635	7.175	9.384	6.993	6.970
Methionine	2.893	3.421	3.419	9.543	4.854	3.543	0.013	1.605	11.193	4.685	2.399	10.506	10.081	3.621	3.010	9.806
Isoleucine	6.234	3.686	3.392	4.921	2.835	3.377	4.345	4.749	2.529	3.904	2.609	2.661	3.493	4.006	3.592	2.443
Leucine	4.702	4.430	5.441	4.554	7.554	6.861	7.018	7.048	6.630	6.829	9.276	6.458	7.634	5.830	7.241	5.821
Tyrosine	3.290	4.874	3.842	3.286	4.336	10.490	2.844	3.353	1.279	2.701	7.394	2.045	2.202	2.984	6.342	2.254
Phenylalanine	5.064	3.897	6.160	4.196	6.394	5.265	5.126	4.332	4.326	4.615	2.854	4.575	3.290	3.597	4.760	3.986
Histidine	2.547	2.837	2.761	2.215	2.054	3.071	2.449	3.524	2.624	2.934	2.688	2.639	2.894	3.137	3.061	2.196
Lysine	4.027	4.044	3.509	2.868	3.425	3.351	4.931	3.266	3.871	5.858	2.509	4.521	4.847	4.844	3.529	3.815
Arginine	5.775	4.948	4.744	4.645	5.392	3.247	4.399	3.986	3.067	2.775	3.667	4.633	5.027	3.748	3.416	3.768

M: Monsoon; PM: Post monsoon; S: Summer; PRM: Pre monsoon.

Table 1.6: Biochemical Composition of the Copepod, *O. brevicornis*, at Station 1 (Bay of Bengal)*

Parameter (%)	*April, 2004*	*May*	*June*	*July*	*Aug.*	*Sep.*
Moisture	81.76±1.41	80.81±0.78	81.96±2.10	81.67±0.81	80.89±1.62	80.84±0.71
Protein	67.77±0.38	66.39±0.75	66.09±2.95	66.66±3.51	64.65±0.89	63.86±1.85
Lipid	15.37±1.66	16.56±1.83	13.81±0.49	14.47±0.25	14.62±0.37	13.97±1.17
Carbohydrate	4.87±0.55	4.76±0.25	3.99±0.45	4.26±0.48	4.73±0.46	4.52±0.51
Ash	3.49±0.20	3.69±0.50	4.13±0.41	3.66±0.20	3.73±0.39	3.98±0.44
Parameter (%)	*Oct., 2004*	*Nov.*	*Dec.*	*Jan., 2005*	*Feb.*	*Mar.*
Moisture	82.60±0.34	83.34±0.34	81.42±1.27	81.45±1.35	81.99±1.68	81.95±1.32
Protein	61.36±1.43	60.37±0.43	60.64±0.12	64.33±0.59	62.92±3.09	64.33±3.48
Lipid	12.59±2.43	12.23±0.55	11.21±0.43	13.75±1.47	13.10±1.44	13.48±0.46
Carbohydrate	3.90±0.51	4.14±0.62	5.70±0.85	5.35±0.44	5.85±0.83	4.07±0.58
Ash	3.76±0.29	3.49±0.22	3.78±0.18	3.62±0.25	3.77±0.54	3.82±0.24

*: Means of triplicate samples ± standard deviation.

Table 1.7: Biochemical Composition of the Copepod, *O. brevicornis*, at Station 2 (Mouth of the Coleroon River)*

Parameter (%)	*April, 2004*	*May*	*June*	*July*	*Aug.*	*Sep.*
Moisture	82.07±2.10	80.78±0.75	81.27±2.75	81.04±1.10	81.54±0.93	81.20±0.48
Protein	66.81±0.33	63.88±1.07	63.57±0.93	64.67±1.20	64.15±0.67	63.87±0.81
Lipid	14.87±0.76	14.54±0.49	15.10±0.44	14.67±1.02	15.14±0.19	15.01±0.33
Carbohydrate	4.57±0.17	4.99±0.63	5.18±0.48	4.57±0.45	4.98±0.50	5.22±0.31
Ash	3.41±0.23	4.29±0.08	4.13±0.45	3.77±0.22	3.62±0.09	3.59±0.18
Parameter (%)	*Oct., 2004*	*Nov.*	*Dec.*	*Jan., 2005*	*Feb.*	*Mar.*
Moisture	81.73±1.51	82.49±0.95	82.64±0.80	82.98±0.44	81.77±1.21	83.25±0.12
Protein	63.99±1.24	61.95±2.19	64.67±3.37	64.24±0.48	63.60±1.22	65.84±1.49
Lipid	11.16±0.38	10.84±0.67	11.88±0.54	13.48±1.11	13.23±1.03	13.94±0.38
Carbohydrate	4.44±0.28	4.06±0.30	4.76±0.27	5.47±0.34	5.33±0.65	4.55±0.46
Ash	3.96±0.23	3.95±0.40	3.93±0.53	3.66±0.05	4.09±0.09	3.73±0.34

*: Means of triplicate samples ± standard deviation.

Table 1.8: Biochemical Composition of the Copepod, *O. brevicornis*, at Station 3 (Coleroon Estuary)*

Parameter (%)	*April, 2004*	*May*	*June*	*July*	*Aug.*	*Sep.*
Moisture	83.05±1.52	81.15±0.88	82.96±2.40	81.77±1.19	82.21±0.72	81.32±0.55
Protein	66.96±0.54	64.38±1.63	64.05±0.52	63.74±1.23	63.89±0.62	63.37±0.75
Lipid	14.29±0.81	14.37±0.34	14.65±1.19	14.21±0.56	15.07±0.22	15.21±0.03
Carbohydrate	4.66±0.31	4.69±0.44	4.93±0.60	4.22±0.60	5.19±0.39	5.03±0.05
Ash	3.40±0.24	4.26±0.08	3.95±0.50	3.71±0.13	3.59±0.13	3.49±0.01
Parameter (%)	*Oct., 2004*	*Nov.*	*Dec.*	*Jan., 2005*	*Feb.*	*Mar.*
Moisture	82.24±1.57	82.69±0.99	82.65±0.80	83.12±0.30	82.84±1.17	83.29±0.17
Protein	62.94±1.41	61.41±2.23	62.68±3.37	64.37±0.26	63.92±1.66	66.65±1.61
Lipid	10.74±0.83	11.61±0.94	11.73±0.66	14.18±0.93	13.99±1.03	13.86±0.30
Carbohydrate	4.44±0.28	4.08±0.30	4.84±0.14	5.44±0.29	5.03±0.47	4.52±0.45
Ash	4.02±0.14	3.82±0.53	3.78±0.44	3.65±0.05	4.16±0.18	3.58±0.22

*: Means of triplicate samples ± standard deviation.

Table 1.9: Biochemical Composition of the Copepod, *O. brevicornis*, at Station 4 (Vettar Backwater)*

Parameter (%)	*April, 2004*	*May*	*June*	*July*	*Aug.*	*Sep.*
Moisture	82.08±2.09	80.87±0.65	81.66±2.60	81.28±0.96	81.19±1.18	81.16±0.54
Protein	67.19±0.91	64.83±2.31	64.23±0.64	66.01±3.24	64.24±0.75	64.39±1.53
Lipid	15.40±1.50	15.45±1.96	14.98±0.64	14.37±0.65	14.82±0.56	14.71±0.84
Carbohydrate	4.41±0.16	4.83±0.47	4.69±0.92	4.48±0.41	4.91±0.59	4.88±0.28
Ash	3.52±0.28	4.03±0.41	4.21±0.49	3.58±0.15	3.83±0.29	3.68±0.32
Parameter (%)	*Oct., 2004*	*Nov.*	*Dec.*	*Jan., 2005*	*Feb.*	*Mar.*
Moisture	82.31±1.04	83.05±0.55	82.51±1.00	82.18±1.52	81.47±1.40	82.57±1.28
Protein	62.28±2.38	61.30±2.28	62.62±3.43	64.32±0.35	61.91±2.23	65.68±1.62
Lipid	12.31±2.02	11.60±0.92	11.85±0.56	13.30±1.35	13.09±1.19	13.58±0.45
Carbohydrate	4.24±0.45	3.90±0.42	5.30±0.69	5.36±0.19	5.38±0.72	4.18±0.57
Ash	3.86±0.39	3.94±0.40	3.70±0.45	3.54±0.17	3.95±0.22	3.74±0.35

*: Means of triplicate samples ± standard deviation.

Table 1.10: Seasonal Variations in Amino Acid Composition of *O. brevicornis* Collected from Coleroon Coastal Waters

Amino Acid	*A. erythraea (Sea)*				*A. erythraea (Mouth)*				*A. erythraea (Estuary)*				*A. erythraea (Vettar Backwaters)*			
	M	*PM*	*S*	*PRM*	*M*	*PM*	*S*	*PRM*	*M*	*PM*	*S*	*PRM*	*M*	*PM*	*S*	*PRM*
Aspartic acid	7.184	7.111	6.724	6.312	7.632	6.443	8.102	8.411	7.655	7.465	6.732	8.010	7.502	7.142	7.016	8.249
Threonine	15.885	13.474	15.311	14.424	15.501	13.280	15.471	14.687	15.137	14.915	17.679	16.584	14.259	14.654	16.943	15.767
Serine	9.413	3.998	8.726	8.599	9.757	8.091	9.167	7.689	10.673	9.561	7.672	10.667	10.465	9.125	7.720	10.437
Glutamic acid	12.834	19.563	13.093	12.495	15.337	14.441	15.594	16.274	13.356	12.221	15.165	14.241	14.236	13.025	15.219	14.378
Glycine	3.971	4.485	3.917	4.216	3.816	4.242	4.265	4.231	4.397	4.217	4.752	4.670	3.957	3.984	4.167	4.489
Alanine	7.331	10.734	7.862	12.107	8.636	10.129	8.318	8.212	8.341	7.474	9.822	8.885	7.891	7.697	9.554	8.596
Valine	7.835	7.631	7.218	10.4	7.603	7.9	7.225	7.393	6.255	6.596	8.686	7.607	5.987	7.561	8.739	7.624
Methionine	6.109	0.979	10.402	0.022	1.027	10.689	0.638	0.593	3.205	11.320	1.753	0.006	2.873	10.732	2.006	0.000
Isoleucine	5.979	2.317	4.325	3.236	2.635	2.035	2.359	2.737	2.939	2.764	3.096	2.814	5.461	2.540	3.547	2.944
Leucine	1.873	8.510	4.746	9.728	7.999	5.230	8.840	9.154	9.012	5.638	8.875	8.548	8.429	5.337	8.495	8.463
Tyrosine	5.148	4.246	2.303	3.142	2.047	2.143	2.375	2.645	2.476	1.958	1.612	2.608	1.736	1.945	1.548	2.649
Phenylalanine	7.161	5.327	4.028	4.980	5.482	5.181	5.479	5.510	4.614	4.371	4.443	5.127	4.245	4.473	4.651	5.142
Histidine	1.971	2.615	2.769	2.599	2.952	3.006	2.181	2.254	2.851	2.498	2.279	1.945	2.478	2.134	2.086	2.053
Lysine	4.132	5.055	4.099	4.463	5.009	3.972	5.097	5.051	4.633	4.278	5.417	4.354	4.310	3.965	5.272	4.273
Arginine	4.038	4.875	4.787	3.910	5.002	3.835	5.791	5.877	5.259	5.623	2.919	4.442	4.928	5.720	2.845	4.387

M: Monsoon; PM: Post monsoon; S: Summer; PRM: Pre monsoon.

protein, lipid, carbohydrate and ash were: 83.83-85.21; 69.42-73.71; 6.63-17.35; 4.22-6.15 and 3.45-4.39 respectively.

Total of 16 amino acids were observed in *A. erythraea*, of these, amino acids like threonine, glutamic acid, alanine, aspartic acid, serine and valine were comparatively higher than other compounds (Table 1.5).

Proximate Composition of *O. brevicornis*

At stn 1, the moisture, ranges (per cent) of protein, lipid, carbohydrate and ash were: 80.81-83.34; 60.37-67.77; 11.21-16.56; 3.90-5.85 and 3.49 to 4.13 respectively. At Station 2, the moisture, ranges (per cent) of protein, lipid, carbohydrate and ash were: 80.78-82.98; 61.95-66.81; 10.84-15.14; 4.06-5.47 and 3.41-4.29 respectively. At station 3, the moisture, ranges (per cent) of protein, lipid, carbohydrate and ash were: 81.15-83.29; 61.41-66.96; 10.74-15.21; 4.08-5.44 and 3.40-4.26 respectively. The proximate composition values of *O. brevicornis* are given in Tables 1.5–1.8. At station 4, the percentages of moisture, protein, lipid, carbohydrate and ash were: 80.87-83.05; 61.30-67.19; 11.60-15.45; 3.90-5.38 and 3.52-4.21 respectively.

In *O. brevicornis*, totally 15 amino acids were recorded. Out of 15, threonine, glutamic acid, serine, alanine, aspartic acid, valine and methionine were observed in higher amount (Table 1.10).

Discussion

Assessment of biochemical composition, such as protein, up to the level of amino acids, lipid and carbohydrate in copepods is important for better understanding of the organic production and cycling of biogeochemical elements in the marine and estuarine biotopes.

Protein formed the major biochemical component in the copepods, *A. erythraea* and *O. brevicornis*. The protein content of the present work was higher than the earlier reports of Sumitra Vijayaragavan *et al.* (1982) in south-central Arabian Sea and Nandakumar *et al.* (1988) in northern part of central Arabian Sea. The protein contents were higher in sea when compared to estuarine mouth and estuary and which indicates that salinity of the water also influences the protein content of the copepods (Raymont, 1972). The protein content of copepods showed great variations, which are comparable with those reported for mixed zooplankton from higher latitudes (Goswami *et al.*, 1981 and Nandakumar *et al.*, 1988). The dominance of protein in *A. erythraea* and *O. brevicornis* may be because of the phytoplanktonic food in wild copepods of Laccadive sea (Rosamma Stephen *et al.*, 1979); Cochin back waters (Madhupratap *et al.*, 1979); Andaman sea (Goswami *et al.*, 1981); northeastern Arabian sea (Krishnakumari and Achuthankutty, 1989); northwest bay of Bengal (Sreepada *et al.*, 1992 and Krishnakumari and Goswami, 1993). Raymont and Conover (1961) observed a higher protein content in the copepod, *Calanus finmarchicus*. Ferguson and Raymont (1968) have reported higher protein values in *Euphasia superba*. Protein formed the major fraction, indicating its usefulness as energy reserve (Conover, 1964 and Conover and Corner, 1968). The observed variations in the protein content might be due to the fact that it is utilizated as a metabolic substrate (Sreepada *et al.*, 1992 and Nageswara Rao and Krupanidhi, 2001). Raymont and Conover (1961) noticed that the protein content was positively related with salinity of the waters. Gopalakrishnan *et al.* (1977) reported the protein content of *Acartia* sp. from Cochin backwaters which were slightly lower than the present values. For coldwater zooplanktonic forms, protein values ranged between 30 and 83 per cent were reported and the highest value was observed earlier in the marine copepod, *A. clausi* (Nakai, 1985).

The lipid content was slightly higher than that of carbohydrate and lower than that of protein. The continuous supply of phytoplankton food would render lipid reserve unnecessary, which might account for the low lipid content in copepod (Madhupratap *et al.*, 1979; Bhat and Wagh 1992; Ingole and Parulekar, 1995 and Nageswara Rao and Krupanidhi, 2001). The lipid content of tropical zooplankton (when compared to temperate zooplankton) is significantly low which might be due to the hydrological conditions and occurrence of food organisms in environment and it was shown by the findings of Sreepada *et al.* (1992). Nageswara Rao and Krupanidhi (2001) proved that the variations in the lipid content also attributed to its storage and utilization during period when it serves as an effective energy reserve. The function of protein as an important energy reserve may be in turn is of importance for zooplankton having low lipid content.

The present results are in agreement with the earlier findings of Lee and Hirota (1973) and Maruthanayagam and Subramanian (1999), suggesting a geographical pattern in the distribution of lipid content between surface water (3 per cent) and deepwater copepods (59 per cent) and concluded that the function of protein as an important energy reserve which may be true for zooplankton having low lipid content, but less important for the lower mesopelagic and bathypelagic copepods which are having high lipid content (Lee and Hirota, 1973 and Maruthanayagam and Subramanian, 1999). Lee (1974) also observed comparatively low lipid content for the ctenophores, *Pleurobrachia pileus* (9 per cent) and *Beroe cucumis* (14 per cent) compared to copepod species, which had a high lipid content (*Calanus finmarchicus*–47 per cent). Conover and Corner (1968) observed the changes in lipid content of the coldwater oceanic copepod *Calanus hyperboreus* to vary from 15 to 50 per cent of the dry weight. Conover (1964) found that the lipid content of this species might increase upto 60 per cent. Gopalakrishnan *et al.* (1977) reported an average lipid content of 18.6 per cent for copepods in general and 25.3 per cent in *Acartia* sp. from the backwaters of Cochin.

In general, carbohydrate content was very low in the copepods when compared to protein and lipid. The lower values of carbohydrate of copepods (wild) have been already reported by (Goswami *et al.*, 1981; Sumitra Vijayaragavan *et al.*, 1982; Krishnakumari and Goswami, 1993; Maruthanayagam and Subramanian, 1999; Nageswara Rao and Krupanidhi, 2001 and Ashok Prabu *et al.*, 2005). The presently observed low carbohydrate content could be attributed to the fact that glycogen is the useful storage carbohydrate in many animals. The utilization of carbohydrate glucosamine during the chitin synthesis in crustaceans may cause decrease of carbohydrate level in copepod (Amsler and George, 1984 and Ashok Prabu *et al.*, 2005). The carbohydrate content was low in all the organisms with a maximum of 5.13 per cent in copepod, *Eucheata marina* and 4.65 per cent in mysids, *Siriella* sp. As in all zooplankton species so far investigated from different marine areas, carbohydrate was the lowest (Nair *et al.*, 1975). Nair *et al.* (1975) reported the carbohydrate content of copepods from Cochin backwater as 3.8 per cent and 2.7 per cent for mysids, which are slightly lower than the presently recorded values. They opined that fluctuations in hydrographic parameters could lead to variations in the carbohydrate content of copepods. Gopalakrishnan *et al.* (1977) reported 4.1 per cent of carbohydrate for *Acartia* sp. from Cochin backwater area, which is higher than the average values (2.54 per cent) reported for copepods in general and less than that for *Euchaeta marina* (5.1 per cent). Even though copepods which feed on phytoplankton could obtain carbohydrate directly, the lower carbohydrate content suggests that glycogen, the usual storage form of carbohydrate, did not form a substantial part of the body reserve (Raymont *et al.*, 1971).

Goswami *et al.* (1981) reported that carbohydrate content of the zooplankton community is dependent upon its composition, declining in gelatinous forms than those with calcareous shells and increasing with copepods. The fluctuations in glycogen content of the animals generally depend upon their feeding activities (Bhat and Wagh, 1992). The low carbohydrate content and the high levels of protein in zooplankton suggest that protein, in addition to lipid, may function as a food reserve

(Conover, 1964). Raymont and Conover (1961) proposed that the carbohydrate from the food may be oxidized directly by zooplankton or that fats may be oxidized on need, or stored as principal reserve food. In general, the observed low carbohydrate content in zooplankton led to contemplation's on the functional role of other biochemical fractions in their metabolism. In the present study, the carbohydrate (5.80 per cent) of *A. erythraea* and *O. brevicornis* content was found to be much higher than those earlier recorded for Indian workers (Goswami *et al.*, 1981; Maruthanayagam and Subramanian 1999; Nageswara Rao and Krupanidhi, 2001 and Ashok Prabu *et al.*, 2005) but quite similar to those recorded by Vinogradov (1964).

During the present analysis, totally 16 and 15 amino acids were observed in wild copepods, *A. erythraea* and *O. brevicornis* respectively with leading of threonine, glutamic acid, alanine, aspartic acid, serine, valine and methionine. The essential amino acids contents were earlier found in freshwater zooplankton like *Daphnia pulex*, *Diatomus* sp., *Cyclops* sp., *Brachionus plicatilis* and in marine diatoms (Yurkuwski and Tabachek, 1979).

Only very limited information is available on the amino acid content of zooplankton. Dabrowski and Rusiecki (1983) have analysed the amino acid content of *Cyclops strenus* and found nine amino acid only. Kibria *et al.* (1997) found 10 amino acids from the cultured *Daphnia* sp., *Brachionus* sp., with huge amount of leucine. Likewise Kokova *et al.* (1990) found 10 amino acids from the cultured *Daphnia* sp. with predominance of lysine. Santhanam and Perumal (2001) found 10 amino acids from the cultured copepod, *O. rigida* with the predominance of norvaline. Rajkumar and Kumaraguru Vasagam (2006) observed 16 amino acids in the cultured copepod, *A. clausi*, with a predominance of lysine, alanine and glutamic acid. Further studies are needed along this line which is important in understanding their physiological functions, metabolism and nutritive value.

From the present study we assume that the two species of copepods are nutritionally equal to any other food fish and they could be exploited successfully for feed purpose and for preparing various by-products similar to fish and prawn by-products or selected animals and may be even used as live feed for aqua farming purpose.

Acknowledgement

The authors are grateful to the Director, CAS in Marine Biology and authorities of Annamalai University for the facilities. Authors are indebted to the renowned experts Dr. Sigrún Jónasdóttir and Dr. P.J.O. Bryen for their editing of the English-language text of an earlier version of the manuscript. The first author (M.R) thanks the Council of Scientific and Industrial Research, (Govt. of India) New Delhi, for the award of a Senior Research Fellowship (Extended).

References

Amsler, M.O.L. and R.Y. George, 1984. *J. Crus. Biol.*, 4: 546-553.

AOAC, 1995. Official methods of analysis, 16th Ed. Association of Official Analytical Chemists. Washington, DC.

Ashok Prabu, V., P. Perumal and M. Rajkumar, 2005. Biochemical composition of some marine copepods. *Res. J. Chem. Environ.*, 9: 36-41.

Bhat, L.K. and A.B. Wagh, 1992. Biochemical composition of zooplankton of Bombay high (oil platform) area in the Arabian sea. *Indian J. Mar. Sci.*, 21: 220-223.

Conover, R.J. 1964. Food relation and nutrition of zooplankton. Univ. Rhode Island. pp. 81.

Conover, R.J. and E.D.S. Corner, 1968. Respiration and nitrogen excretion by some marine zooplankton in relation to their life cycles. *J. Mar. Biol. Ass. U.K.*, 48: 49-75.

Dabrowski, D. and M. Rusiecki, 1983. Content of total and free amino acid in zooplanktonic food of fish larvae. *Aquaculture*, 30: 31-42.

Dubois, M., K.A. Gills, J.K. Hamilton, P.A. Rober and F. Smith, 1956. Colorimetric method for determination of sugars and related substances. *Anal. Chem.*, 28, 350 pp.

Ferguson, C.F. and J.J.B. Raymont, 1968. Biochemical studies on marine zooplankton. XII. Further investigations on *Euphausia superba* Dana. *J. Mar. Biol. Ass. U.K.*, 54: 719-725.

Finlayson, A.J. 1964. Amino acid recovering in the analysis of some feed samples. *Can. J. Plant Sci.*, 45: 184-188.

Finn, R.N., H.J. Fyhn and M.S. Evjen, 1995. Physiological energetics of developing embryos and yolk-sac larvae of Atlantic cod (*Gadus morhua*). I. Respiration and nitrogen metabolism. *Mar. Biol.*, 124: 355-369.

Fluchter, J. and H. Rembold, 1986. Soluble factor essential for metamorphosis of coreconid larvae has been partially purified from *Artemia salina*. *Arch. Hydrobiology*, 22: 197-202.

Folch, J.M., M. Lees and G.H. Sloane-Stanley, 1956. A simple method for the isolation and purification of total lipids from animal tissues. *J. Biol. Chem.*, 226: 497-509.

Gopalakrishnan, T.C., K.K.C. Nair, M. George Peter, V.N. Sankaranarayanan and T.S.S. Rao, 1977. Major Biochemical constituents of some faunal components of the Cochin Backwaters. *Indian J. Mar. Sci.*, 6: 151-153.

Goswami, S.C., L. Krishnakumari and Y. Shrivastava, 2000. Diel variations in zooplankton and their biochemical composition from Vengurla to Ratnagiri, West coast of India. *Indian J. Mar. Sci.*, 29: 277-280.

Goswami, S.C., T.S.S. Rao and S.G.P. Matondkar, 1981. Biochemical composition of zooplankton from the Andaman sea. *Indian J. Mar. Sci.*, 10: 296-300.

Holm, J.C. and D. Moller, 1984. Growth and prey selection by Atlantic salmon yearlings reared on live freshwater zooplankton. *Aquaculture*, 43: 401-412.

Ingole B.S. and A.H. Parulekar, 1995. Biochemical composition of Antarctic zooplankton from the Indian Ocean sector. *Indian J. Mar. Sci.*, 24: 73-76.

Kasturirangan, L.R. 1963. A Key for the identification of the more common planktonic copepoda of Indian coastal waters. Publication No 2, CSIR. Publication, 87 pp.

Kibria, G., D. Nugegoda, R. Fairclough, P. Lam and A. Bradley, 1997. Zooplankton: Its biochemistry and significance in aquaculture. *Naga*, ICLARM, 8-14.

Kokova, V.Ye., V.A. Baraskov and I.N. Trubachev, 1990. Free amino acids of certain aquatic invertebrates. *Hydrobiologia*, 26: 77-79.

Krishnakumari K. and S.C. Goswami 1993. Biomass and biochemical composition of zooplankton from Northeast Bay of Bengal during January 1990. *Indian J. Mar. Sci.*, 22: 143-145.

Krishnakumari, L. and C.T. Achuthankutty, 1989. Standing stock and biochemical composition of zooplankton in the northeastern Arabian sea. *Indian J. Mar. Sci.*, 18, 103-105.

Krishnakumari, L. and R. Vijayalakshmi Nair, 1988. Biomass, organic carbon and calorific content of zooplankton from the Arabian sea off Central west coast of India. *Indian J. Mar. Sci.*, 17: 78-80.

Lauff, M. and R. Hofer, 1984. Proteolytic enzymes in fish development and the importance of dietary enzymes. *Aquaculture*, 37: 335-346.

Lee, F.R. 1974. Lipids of zooplankton from Bute Inlet, British Columbia. *J. Fish. Res. Bd. Can.*, 31: 1577-1582.

Lee, R.F. and J. Hirota, 1973. Wax esters in tropical zooplankton and nekton and the geographical distribution of wax esters in marine copepods. *Limnol. Oceanogr.*, 18: 227-239.

Lemm, C.A. 1983. Growth and survival of Atlantic salmon fed semi-moist or dry starter diets. *Prog. Fish. Cult.*, 45: 72-75.

Madhupratap, M., P. Venugopal and P. Haridass, 1979. Biochemical studies on some tropical esturine zooplankton. *Indian J. Mar. Sci.*, 8: 155-157.

Maruthanayagam, C. and P. Subramanian, 1999. Biochemical variation of zooplankton population. *J. Mar. Biol. Ass. India*, 41: 111-115.

Mearns, K.J. 1986. Sensitivity of brown trout (*Salmo trotta* L.) fry to amino acids at the start of exogenous feeding. *Aquaculture*, 55: 191-200.

Millamena, O.M., V.D. Penaflorida and P.F. Subosa, 1990. The macronutrient composition of natural food organisms mass cultured as larval food for fish and prawns. *Israeli. J. Aqua. Bamidgeh*, 42: 77-83.

Munilla-Moran, R., J.R. Stark and A. Barbour, 1990. The role of exogenous enzymes in digestion in cultured turbot larvae (*Scophthalmus maximus* L.). *Aquaculture*, 88: 337-350.

Nageswara Rao, I. and G. Krupanidhi, 2001. Biochemical composition of zooplankton from the Andaman sea, *J. Mar. Biol. Ass., India*, 43: 49-56.

Nair, V.R., M. Saraswathy, K. Sarala Devi and T.C. Gopalakrishnan, 1975. *Mahasagar-Bull. Natn. Inst. Oceanogr.*, 8: 109.

Nakai Z. 1985. *Spec. Publs. Tokai. Fish. Res. Lab*, 5: 12.

Nandakumar, K., L.K. Bhat and A.B. Wagh, 1988. Biochemical composition and calorific value of zooplankton from Northern part of Central Arabian sea. *Indian J. Mar. Sci.*, 17: 48-50.

Rajkumar, M. and K.P. Kumaraguru vasagam, 2006. Suitability of the copepod, *Acartia clausi* as a live feed for Seabass larvae (*Lates calcarifer* Bloch): Compared to traditional live-food organisms with special emphasis on the nutritional value. *Aquaculture*, 261: 649-658.

Raymont J.*E.G.* and R.J. Conover, 1961. Further investigations on the carbohydrate content of marine zooplankton. *Limnol. Oceanogr.*, 6: 154-164.

Raymont J.*E.G.*, R.T. Srinivasagam and J.K.B. Raymont, 1971. Biochemical studies on marine zooplankton. VIII. Further investigations. *Deep Sea Res.*, 18: 1167-1178.

Raymont, J.*E.G.* 1972. Essays in Hydrobiology. University of Exetter, Southampton, UK. pp. 83.

Rosamma Stephen, S.V. Panampunnayil, T.C. Gopalakrishnan and V.N. Sankara Narayanan, 1979. Biochemical studies on zooplankton from the Laccadive sea (Lakshadweep). *Indian J. Mar. Sci.*, 8: 257-258.

Santhanam, P. and P. Perumal, 2001. Note on the amino acid profile of cultured copepod, *Oithona rigida* Giesbrecht. *Adv. Bios.*, 20: 83-88.

Schipp, G.P., J.M.P. Bosmans and A.J. Marshall, 1999. A method for hatchery culture of tropical calanoid copepoda, *Acartia* spp. *Aquaculture*, 174: 81-88.

Spackman, D.H., W.H. Stein and S. Moore, 1958. Automatic recording apparatus for use in the chromatography of amino acids. *Anal. Chem.*, 30: 1190-1195.

Sreepada, R.A., C.U. Rivonker and A.H. Parulekar, 1992. Biochemical composition and calorific potential of zooplankton from Bay of Bengal. *Indian J. Mar. Sci.*, 21: 70-73.

Stottrup J.G. 2000. The elusive copepods: Their production and suitability in marine aquaculture. *Aqua. Res.*, 31: 703-711.

Sumitra Vijayaragavan, R.A. Selvakumar and T.S.S. Rao, 1982. Studies on zooplankton from the Arabian sea off the South- Central west coast of India. *Indian J. Mar. Sci.*, 11: 70-74.

Szyper, J.P. 1989. Nutritional depletion of the aquaculture feeding organisms: Spp. *Euterpina acutifrons*, *Artemia* and *Brachinus plicatilis* during starvation. *J. World Aqua. Soc.*, 20: 162-169.

Thakur, N.K. 1998. Importance of live food organisms in aqua hatcheries. Training manual on culture of live food organisms for aqua hatcheries, CIFE, Deemed University, ICAR, Mumbai, pp. 1-4.

Tucker, J.W. 1992. Feeding intensively-cultured marine fish larvae. In: Proc. Aquaculture Nutrition Workshop, Salamander Bay, 15-17 April 1991. NSW Fisheries, Brackishwater Fish Culture Research Station, Salamander Bay, Australia, pp. 129-146.

Vinogradova Z.A. 1964. Some biochemical aspects of a comparative study of plankton from the Black sea, the sea of Azov and the Caspian sea. *Okenologiya*, 4: 232-242.

Yurkowski, M. and J.L. Tabachek, 1979. Proximate and amino acid composition of some natural fish foods. In. J.E. of the world symposium on finfish. Nutrition and fish feed technology, Hamburg, 20-23 June, 1978 Berth Vol. 2, 435-448.

Chapter 2

Studies on the Toxic Effects of Arsenic on Glucose Content in Freshwater Fish, *Labeo rohita* (Hamilton)

K. Pazhanisamy

Introduction

India is a developing country and it has tremendously advanced towards industrial development. This has resulted, in the environmental pollution, which includes atmosphere, lithosphere and hydrosphere. Pollution in an undesirable change in the physical, chemical, or biological characteristics of our air, land and water that may or will affect the life of the human beings as well as other living organism (Odum, 1971). Now-a-days, environmental pollution by toxic heavy metals, chemicals and pesticides is widespread and has become a global problem. These toxic chemicals and metals have changed the quality of waters that affect the fish along with other aquatic organism (Dhasarathan *et al.*, 2000).

Toxicology is one of the most important multidisciplinary subject in animal and plant sciences dealing with the adverse effects of chemicals and allied agents on living system. It is also valuable in the protection of public health against hazards associated with toxic substances in food, air and water (Lu, 1985). During the past few decades, rising trends of population growth, development of modern technology, heavy industrialization and modernized agricultural practices have brought a dramatic increase in production and consumption of a large variety of new synthetic chemicals, and there by high amount of pollutants is in to discharged the aquatic environment, of which heavy metals play a major role.

Saikia *et al.* (1988) have pointed out that most of the heavy metals are toxic to organisms as well as human beings of the exposure levels are relatively high. Among the various heavy metals, arsenic is one of the most persistent metal in the environment. It poses threat to the aquatic species and has a detrimental effect on the aquatic organisms, especially fishes (Pazhanisamy, 2002). Carbohydrate is an important energy source of all vital activities of an organism. Glucose a reliable and immediate source of energy is present in almost all tissues. The glucose is the most important fraction of carbohydrate and it serves as the major metabolic fuel.

Baskaran *et al.* (1989) have observed a decrease in carbohydrate reserves in liver and muscle of *Oreochromis mossambicus* when exposed to textile mill effluent. A decrease in the liver glucose content has been reported for *Heteropneustes fossilis* exposed to aldrin (Singh and Srivastava, 1992) and zinc (Sastry and Subhadra, 1983).The level of liver and muscle glucose declined when the fish, *Channa punctatus* was treated with phenyl mercuric acetate (Karuppasamy, 1999). Similarly, decrease in liver and muscle glucose level has been reported in the *Oreochromis mossambicus* when exposed to ethofenprox (Muniyan, 1999) and in *Mystus vittatus* exposed to copper (Rajamanickam, 1992). But, out of these studies, adequate information is not available on the effect of arsenic on glucose content of the fish. Hence, an attempt has been made in the present study, to examine the lower and higher sublethal concentration of arsenic and glucose content in liver and muscle tissues of the fish *Labeo rohita* exposed to different time intervals on 7,14,21 and 28 days.

Materials and Methods

Freshwater fish, *Labeo rohita* ranging from 10 to 12 cm in length and weighing between 9 to 14 g were collected from fish farm located in Puthur, 10 km away from university campus and were acclimated under laboratory conditions (29±1° C). The fish were fed daily on oil less groundnut cake. The unused food was renewed after 2 hours and the water was changed daily. Prior to experimentation *Labeo rohita* were acclimated in experimental tanks for at least one week. The LC_{50} for arsenic trioxide for 96 hours was determined according to Finney (1971) and $1/10^{th}$ of the LC_{50} value (0.27ppm) and $1/3^{rd}$ of the LC_{50} value (0.91ppm) were taken as lower and higher sublethal concentrations respectively. The fish were exposed to these concentrations for treated and control period of 7, 14, 21 and 28 days. A control group was maintained at an identical environment. The fish were sacrificed from, both experimental and control groups on 7th, 14th, 21st and 28th days of exposure periods and subjected to analysis of the biochemical changes. The glucose content of the tissues sampled from liver and muscle, was estimated by the method of Kemp and Andrienne Kits Van Heijininger (1954).

Results and Discussion

The glucose content in the liver and muscle tissues of control and arsenic treated fish, *Labeo rohita* are depicted in Table 2.1. In the present investigation a significant decrease in glucose level of the liver and muscle of *Labeo rohita* at lower and higher sublethal concentration of arsenic trioxide were evident. The decrease in mean glucose levels were statistically significant ($P < 0.05$) in both sublethal concentrations at all the exposure periods (Table 2.1.)

In the present investigation, the decrease level of glucose was noticed in the liver tissues of *L.rohita* at all exposure periods of arsenic intoxication which suggests that the glucose molecules were utilized to meet the energy demand during stress period. Further, the decrease in the liver glucose content suggests that the rapid transport of a considerable part of glucose molecules from has occurred the liver to muscle via blood to meet extra energy demand for muscular activity. A similar decrease in the liver glucose content has been reported in *Channa punctatus* exposed to phenyl mercuri acceate (Karuppasamy, 1999). Das and Konar (1990) have also reported that the level of liver glucose declined when the fish *Tilapia mossambica* was treated with light crude oil. They have also suggested that the decrease in glucose content in the liver might be due to its transport to the muscle via blood to meet the energy requisites for muscular activity. Such results were also obtained in *Heteropneustes fossilis* exposed to aldrin (Srivastava and Singh, 1981) and in *Puntius stigma* exposed to agrofen and thiodon (Khillare, 1990) which extend support to the present observations.

Table 2.1: Effects of Lower and Higher Sublethal Concentrations of Arsenic on Glucose Levels in Liver and Muscle Tissues of *Labeo rohita* at Different Periods of Exposure

Tissues	*Group*	*Exposure Period in Days*				*F Value*
Liver	C	10.59±0.079	10.61±0.387	10.59±0.095	10.60±0.047	0.019NS
	LC	8.43±0.041 (−20.40)	7.30±0.026 (−31.19)	5.87±0.046 (−44.57)	4.80±0.006 (−54.71)	223.95*
	HC	7.29±0.101 (−31.16)	6.22±0.054 (−41.38)	4.74±0.003 (−55.24)	3.61±0.018 (−65.94)	772.33*
Muscle	C	2.99±0.0077	2.98±0.0047	2.99±0.0121	2.98±0.016	0.040NS
	LC	2.61±0.008 (−12.70)	2.55±0.011 (−14.43)	2.40±0.003 (−19.73)	2.21±0.010 (−25.83)	666.1*
	HC	2.52±0.010 (−15.720)	2.43±0.008 (−18.46)	2.18±0.004 (−27.09)	2.00±0.006 (−32.88)	105.33*

C: Control; LC: Lower sublethal concentration; HC: Higher sublethal concentration; NS: Non significant;
*: Indicates significant at 5 per cent level of F test.

Mean±S.E. indicates the mean of six individual observations.

(+/-) indicates the per cent increase / decrease over the control.

Values expressed in mg/g wet wt of tissue.

In the present investigation, the decreased level of glucose was also noticed in the muscle tissue of *Labeo rohita* at all exposure periods of arsenic exposure which suggest that the rapid utilization of glucose through glycolytic mechanism, possibly leads to the supply of more amount of energy during stress period. Similarly decrease in muscle glucose level has been reported in the *Mystus vittatus* exposed to cadmium (Rengarajan, 1989) and in *Anguilla anguilla* exposed to lindane (Ferrando and Andreu-Moliner, 1990). This is in accordance with the observation of Jagadeesan, (1994) on *Labeo rohita* exposed to mercuric chloride.

Summary

In the present study, the level of glucose in the liver and muscle tissues of control and arsenic treated fresh water fish, *Labeo rohita* have shown remarkable changes. Glucose content in liver and muscle tissues under two different sublethal concentration of arsenic toxicity study were found to decrease continuously through out experimental periods. The tissues–specific depletion of glucose as observed in the present study might be due to its rapid utilization through glycolytic mechanism to meet the energy demand under stress period.

Acknowledgement

The author is greatly indebted to the Department of Zoology, Annamalai University for providing necessary research facilities.

References

Baskaran, P., Palanichamy. S and Arunachalam. S 1989. Effects of textile dye effluent on feeding energetics, body composition and oxygen consumption of the freshwater fish, *Oreochromis mossambicus, J. Ecobiol.*, 1(3): 203-214.

Das, P.K.M.K and Konar, S.L., 1990 Sublethal effects of mixture of light Iranian crude oil and cryptoanionic detergent sandopam DTC on fish *Tilapia mossambica. Environ. Ecol.* 8 (1): 201–207.

Dhasarathan, P., Palaniappan, R. and Ranjit Singh, A.J.A. 2000. Effect of endosulfan and butachlor on the digestive enzyme and proximate composition of the fish, *Cyprinus carpio. Indian J.Environ, and Ecoplan.*, 3(3): 611-614.

Ferrando, M.D and Andreu–Moliner.E 1990. Effects of lindane on fish carbohydrate metabolism. *Ecotoxi Environ. Safe.* 22: 17-23.

Finney, D.J. 1971. Probit Analysis. University Press, Cambridge, pp. 333.

Jagadeesan, G. 1994. Studies on the toxic effects of mercuric choride and influence of antidote dimercaprol on selected tissues in *Labeo rohita* (Hamilton) fingerlings. Ph.D Thesis, Annamalai University, India.

Karuppasamy, R.1999. The effect of Phenyl Mercuric Acetate (PMA) on the physiology, biochemistry and histology of selected organs in a freshwater fish, *Channa punctatus* (Bloch).Ph.D Thesis, Annamalai University, India.

Kemp, A. and Andrienne Kits Van Heijininger, J.M. 1954. A colorimetric method for the determination of glucose and glycogen in tissues. *Biochem.* J.,56: 646-648.

Khillare, Y.K. 1990. Effect of pesticides agrofen and thiodon on the glycogen content of freshwater fish, puntius stigma. *Proc. Int. Symp. Environ. Impect Biosystem Madras, Trends in Ecotoxicology* (1): 43-45.

Lu, 1985. Basic toxicology fundamental, target organs and risk assessment. Hemisphere Publishing Corporation, Washington.

Muniyan, M. 1999. Effects of ethofenprox (trepon) on the biochemical and histological changes in selected organs of the freshwater fish, *Oreochromis mossambicus* (Peters). Ph.D.Thesis, Annamalai University, India.

Odum, E.P., 1971. 3rd Edition, Fundamentals of Ecology, W.B. Saunder's Company.

Pazhanisamy,K. 2002 Studies on the impact of arsenic on a freshwater fish, *Labeo rohita* (Hamilton) Ph.D.,Thesis Annamalai University, India.

Rajamanickam, C. 1992. Effects of heavy metal copper on the biochemical contents,bioaccumulation and histology of the selected organs in the freshwater fish, *Mystus vittatus* (Bloch). Ph.D. Thesis, Annamalai University, India.

Rengarajan, K.V., 1989. Biochemical studies on the effects of cadmium chloride in the freshwater fish *Mystus vittatus* (Bloch). M.Phil., Thesis Annamalai University.

Saikia, D.K., Mathur, R.P. and Srivastava, S.K. 1988. Heavy metals in water and sediments of upper Ganges. *Indian J.of Environ. Hlth.*, 31(1): 11-17.

Sastry, K.V. and Subhadra, K. 1983 Metabolic changes in the freshwater catfish *Heteropneustes fossilis* chemically exposed to zinc. Fourth All India Seminar on Ichthyology. D.A.V. College, Dehradun, India.

Singh, N.N and Srivastava, A.K. 1992. Effect of aldrin on some biochemical parameters of Indian cat fish, *Heteropneustes fossilis. J.Freshwater Biol.* 4 (4): 289–293.

Srivastava, A.K and Singh, N.N., 1981. Effect of aldrin on carbohydrate metabolism in Indian cat fish. *Acta. Pharmacol, Toxicol* 49 (4) 266-269.

Chapter 3

Microbial Resources of the Mangrove and Vellar Estuary, Southeast Coast of India

Surajit Das, P.S. Lyla and S. Ajmal Khan

Introduction

Mangroves are specially adapted group of woody plants that occur in the interphase between land and sea (Kathiresan and Bingham, 2001). The plants usually grow at the intertidal zones of sheltered shores, estuaries, tidal creeks, backwaters, lagoons, marshes and mud-flats in the tropical and sub-tropical latitudes and support genetically diverse groups of aquatic and terrestrial organisms. The mangrove ecosystems are of great ecological and economic significance in coastal production, enrichment of water quality and in production of fishery resources (Odum and Heald, 1975). It is one of the most productive ecosystems on the earth and harbours a wide variety of microorganisms with ecological and economical significance. The dynamic environment of estuary and mangrove ecosystems provide the ideal habitat for different groups of microorganisms which are nourished by detritus and in turn thus help the mangrove ecosystem in different ways (Holguin *et al.*, 2001; Das *et al.*, 2004).

Microbes especially bacteria form an important component of this ecosystem and they contribute towards its steady state by their twin role as mineralisers and producers.

Mangrove ecosystem generates a large amount of litter in the form of leaves, branches, twigs and other debris. The leaves and wood are made mainly of lignocellulose components that are degradable by microorganisms (Alongi *et al.*, 1989) and the degradation starts immediately after its colonization by microorganisms. This is a reversible process but the productivity primarily depends on the microbial resources of the mangrove and estuary. Hence, a periodical monitoring of microbial resources of mangrove and estuary was carried out in the present study to understand the seasonal abundance of microbial population and their generic composition. The microorganisms, besides their role in the productivity have importance in biotechnological search (Shome *et al.*, 1995).

Materials and Methods

The Vellar estuary is situated in the southeast coast of India (Lat. 11°29' N and Long. 79°46' E) (Figure 3.1) and the mangrove has been developed in front of the Centre of Advanced Study in Marine Biology, Annamalai University. This developed mangrove forest covers an area of 2 hectares in the intertidal area. In this forest *Rhizophora mucronata* is abundant followed by *R. apiculata, Avicennia marina, A. officinalis, Aegiceras corniculatum, Ceriops decandra, Bruguiera cylindrica, B. gymnorrhiza, Excocaria agallocha* and *Acanthus ilicifolius*. Sediment Samples were collected from April 2004 to March 2005 seasonally (April to June- Summer; July to September- Pre-monsoon; October to December- Monsoon and January to March- Post-monsoon) from the ten stations equidistantly situated.

Sediment samples were collected from mangrove area and the adjacent estuarine area at a distance of 10m from mangrove by alcohol rinsed Petersen Grab. Sediment sub-samples were colleted from the grab and brought to the laboratory for Total Heterotrophic Bacterial (THB) and actinomycetes population enumeration. Sediment samples were diluted serially and aliquots of samples were used for plating. THB was enumerated in triplicate on Zobell's Marine Agar medium (Peptone: 5.0g; Yeast extract: 1.0g; Ferric citrate: 0.1g; K_2HPO_4: 0.5g; pH: 7.5±0.3; Agar: 15.0g and 50 per cent sea water: 1000ml) by spread plate method (Schneider and Rheinheimer, 1988). Plates were then incubated at 28°C for 2 days and the colonies were counted and expressed in CFU g^{-1} dry wt. sediment. Bacterial

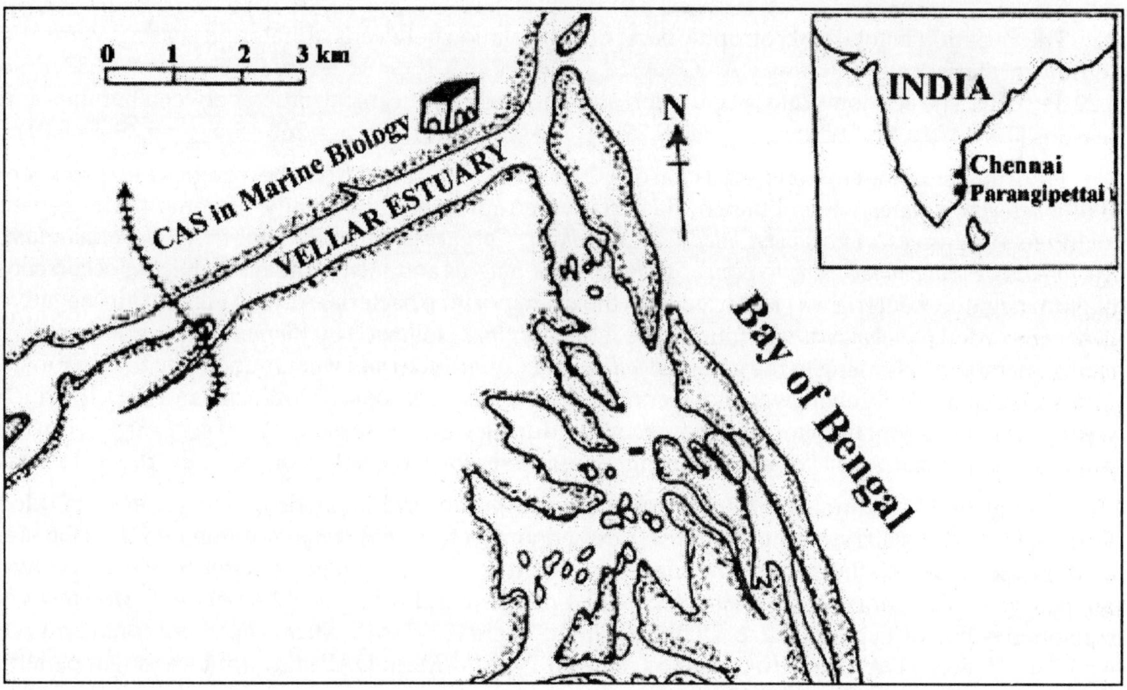

Figure 3.1: Study Area-Vellar Estuary and the Adjacent Mangrove

colonies were purified by successive restreaking and identified upto species level. Identification was carried out performing biochemical tests, *e.g.*, gram staining, glucose utilization, motility, fermentation and antibiotic sensitivity and following the scheme of Simidu and Aiso (1962), Baumann *et al.* (1972), Starr *et al.* (1981), Oliver (1982) and Holt *et al.* (1994).

For actinomycetes population, Starch Casein Agar (SCA) medium (Soluble starch: 10.0g; Casein: 1.0g; pH: 7±0.1; Agar: 15.0g and sea water: 1000ml) was used supplementing nystatin and cyclohexamide at 25 $\mu g\ ml^{-1}$ and 10 $\mu g\ ml^{-1}$ respectively (to minimize contamination with fungi) and 10 µg nalidixic acid ml^{-1} (to minimize bacterial growth) (Ravel *et al.*, 1998; Das *et al.*, 2005). Plates were then incubated at 28°C for 2-4 weeks and the powdery colonies were counted and expressed in CFU g^{-1} dry wt. sediment. Actinomycetes colonies were purified by successive restreaking and identified upto genus level by aerial mass colour, spore chain morphology and cell wall chemotaxonomy (cell wall amino acid and sugar pattern, Lechevalier and Lechevalier, 1970).

Results

The present observation on the total heterotrophic bacterial population and actinomycetes population revealed higher level in the mangrove habitat than those of estuarine habitat. Higher microbial population was observed during the monsoon season. Actinomycetes population was found to be lower than the total heterotrophic bacterial population. Bacteria population in estuarine habitat ranged between 10^4 and 10^6 CFU g^{-1} dry wt. sediment and in mangrove habitats it ranged between 10^5 and 10^7 CFU g^{-1} dry wt. sediment. Actinomycetes population narrowly varied in estuarine habitat (with in 10^3 CFU g^{-1} dry wt. sediment) but widely varied in mangrove habitat (10^2 to 10^4 CFU g^{-1} dry wt. sediment) (Table 3.1).

The variation in total heterotrophic bacterial population between habitats and seasons were also significant statistically (Two-way ANOVA; $F_{Habitats} = 1603.83$, $F_{crit} = 8.53$, $P<0.01$; $F_{Seasons} = 537.37$, $F_{crit} = 5.29$, $P<0.01$). The actinomycetes population also showed significant variation between habitats and seasons (Two-way ANOVA; $F_{Habitats} = 960.29$, $F_{crit} = 8.53$, $P<0.01$; $F_{Seasons} = 266.45$, $F_{crit} = 5.29$, $P<0.01$).

From the estuarine environment a total of 852 bacterial isolates and from the mangrove environment 1008 bacterial isolates were obtained. Bacteria were represented by totally 13 genera. The genera included *Alcaligenes, Aeromonas, Bacillus, Cytophaga, Corynebacterium, Enterobacter, Flavobacterium, Micrococcus, Pseudomonas, Streptococcus, Staphylococcus, Shigella* and *Vibrio* (Table 3.2). Higher occurrence of gram negative bacteria was recorded than the gram positive bacteria. Among eight gram-negative genera recorded *Pseudomonas* was found to be more dominant followed by *Vibrio*. Among gram positive forms, members belonging to the genera *Bacillus* and *Corynebacterium* were found to be the dominant forms. Some bacterial genera were not recorded during some seasons. *Enterobacter* and *Shigella* which were mostly fecal contaminant were not recorded during summer season. *Cytophaga, Shigella* during pre-monsoon season and *Micrococcus* during monsoon season were not recorded in the present study.

A total of 155 actinomycetes isolates were recovered and identified. The genera included *Streptomyces, Micromonospora* and *Nocardia* (rare actinomycetes). *Streptomyces* dominated than the rare actinomycetes in both the estuary and mangrove habitat. The percentage occurrence was as follows: *Streptomyces*- 67.8 per cent, *Micromonospora*- 19.4 per cent and *Nocardia*- 12.8 per cent. *Streptomyces* contained cell wall Type I- LL 2, 6, Diamino Pimelic Acid (LL-DAP). *Micromonospora* contained cell wall Type II- Meso-DAP *Nocardia* contained cell wall Type IV- Meso-DAP and Arabinose sugar pattern.

Table 3.1: Microbial Population (mean±SD, n= 30) in Both the Habitats During Four Seasons

Microbial Resources	Summer		Pre-Monsoon		Monsoon		Post-Monsoon	
	Est.	Mang.	Est.	Mang.	Est.	Mang.	Est.	Mang.
THB (CFU g^{-1} dry sed. wt)	$4.3 \pm 0.12 \times 10^4$	$2.3 \pm 0.02 \times 10^5$	$8.4 \pm 0.02 \times 10^5$	$5.8 \pm 0.44 \times 10^6$	$3.8 \pm 1.12 \times 10^6$	$6.5 \pm 0.53 \times 10^7$	$2.1 \pm 1.5 \times 10^6$	$6.1 \pm 0.42 \times 10^7$
Actino popul (CFU g^{-1} dry sed. wt)	$6.3 \pm 1.2 \times 10^3$	$2.6 \pm 0.8 \times 10^2$	$6.8 \pm 1.3 \times 10^3$	$2.1 \pm 0.3 \times 10^4$	$7.1 \pm 0.05 \times 10^3$	$5.4 \pm 0.51 \times 10^4$	$7.0 \pm 1.6 \times 10^3$	$4.7 \pm 1.8 \times 10^4$

Table 3.2: Percentage (Per cent) Occurrence of Bacterial Genera During Different Seasons

Bacterial genera	*Summer*		*Pre-Monsoon*		*Monsoon*		*Post-Monsoon*		*Total*
	Est.	*Mang*	*Est*	*Mang*	*Est*	*Mang*	*Est*	*Mang*	
Pseudomonas	22.2	20.0	28.4	18.6	25.0	12.6	18.8	25.6	171.2
Vibrio	31.3	25.5	23.5	22.3	10.2	8.5	8.5	9.6	139.4
Alcaligenes	2.0	14.5	20.4	25.5	8.5	30.3	2.5	10.5	114.2
Enterobacter	0	1.5	0	2.3	10.6	12.3	8.5	11.0	46.2
Shigella	0	1.5	1.5	0	8.6	5.3	1.5	3.0	21.4
Aeromonas	3.5	2.5	4.0	3.5	5.5	5.0	8.0	3.0	35.0
Cytophaga	2.0	3.1	1.5	0	3.0	7.4	0	1.5	18.5
Flavobacterium	3.5	2.5	5.7	6.3	7.0	2.5	6.0	10.0	43.5
Bacillus	12	18	8.4	10	8.6	4.0	14.3	5.5	80.8
Micrococcus	4.5	2.5	2.5	4.5	1.5	0	12.3	8.5	36.3
Streptococcus	3.0	2.0	0	3.0	1.5	5.6	3.6	3.6	22.3
Staphylococcus	2.8	3.0	1.5	2.5	5.0	3.5	5.5	4.7	28.5
Corynebacterium	13.2	3.4	2.6	1.5	5.0	3.0	10.5	3.5	42.7

Discussion

The distribution of microorganisms in any aquatic environment is governed by biotic and abiotic factors. In the present study, bacterial population was found to be dominant followed by actinomycetes. However, higher population of both bacteria and actinomycetes was noticed in the mangrove area than those of estuary which is attributed to availability of nutrients and particulate carbon in the mangrove environment due to leaf litter decomposition. The similar trend was also noticed by Chandran and Ramamoorthi (1984) and Kannan and Vasantha (1986). Further, an increase in population density during monsoon season may also be attributed to the addition of terrigenous material through land run-off, which carries higher microbial load as was reported earlier from Vellar estuary (Purushothaman, 1999). The total viable heterotrophic bacteria in an environment depend on the availability of growth supporting organic matter and micronutrients brought from the run-off. The degradation of mangrove vegetative material produces detritus which is rich in energy nourishes microbial populations and helps in nutrient regeneration through nutrient recycling. Hence, microbial activity is responsible for major nutrient transformation (Alongi, 1988).

Seasonal variations and ranges of total heterotrophic bacterial CFUs from sediment samples of estuary and mangrove are at least two orders of magnitude higher than those reported by Lakshmanaperumalsamy *et al.* (1986) and Purushothaman (1999) from Vellar estuary and Prema *et al.* (2005) from Rajakkamargalam estuary.

The generic composition of bacteria observed in the present study is in conformity with the general findings of occurrence of more gram-negative forms compared to the gram-positive forms in the coastal environment. The available reports on the total heterotrophic bacterial population in the coastal waters of India by Kannan and Vasantha, (1986); Prabu *et al.* (1991); Sathiyamurthy and Purushothaman (1992) and Ramaiah *et al.* (1996) are in good agreement with the present observation.

However, the total number of bacterial genera was quite higher than the reports of Vasantha and Kannan (1987) and Purushothaman (1999).

Pseudomonas was found to be dominant in the present study which contradicts with the reports of Kannan and Vasantha (1986), Vasantha and Kannan (1987) and Ramaiah *et al.* (1996) where *Vibrio* was the dominant genus. However, simple sugars, amino acids released with root lysate of mangrove plants promote the growth of pseudomonads (Purushothaman, 1999). Some bacterial genera were not recorded during summer season *e.g.*, *Enterobacter* and *Shigella* which were mostly fecal contaminants. Ramaiah *et al.* (1996) also reported the absence of certain genera during a particular season from Tamil Nadu coastal area.

Actinomycetes are slow growing filamentous gram-positive microbes having the intermediate characteristics of both bacteria and fungi (Cross, 1982). In the mangrove environment actinomycetes diversity is higher and shows promising biotechnological activity (Das *et al.*, 2006). Similar to bacterial population, actinomycetes population also showed higher abundance in the present study than those reported by Walker and Colwell (1975) from Chesapeake Bay, Lakshmanaperumalsamy *et al.* (1986) and Siva Kumar (2001) from Vellar estuary. Such variation is quite significant which indicate the microbial wealth of Vellar estuary and adjacent mangrove environment. The genus *Streptomyces* is the most widely studied and well-known genus of the Actinomycete family. Streptomycetes usually inhabit soil and are important decomposers. They are able to metabolize many different compounds including sugars, alcohols, amino acids and aromatic compounds by producing extracellular hydrolytic enzymes. Their metabolic diversity is due to their extremely large genome which has hundreds of transcription factors that control gene expression, allowing them to respond to specific needs. Thus, different groups of microorganisms contribute to the fertility and productivity of mangrove environment through litter decomposition, removal of H_2S, and synthesis of amino acids, vitamins and other plant growth regulators which ultimately enrich the coastal ecosystem and in turn the fisheries.

Conclusion

The present study indicates that the mangrove environment harbours higher microbial load than the estuary. Monsoon season showed higher abundance of microbial load than summer season. Totally thirteen bacterial genera were recorded. These generic compositions suggest that the existing bacterial communities are likely to play a very active role in the rapid *in situ* degradative processes. Thus, both the habitats are good sources of actinomycetes which are biotechnologically important for secondary metabolite production.

Acknowledgement

Authors are thankful to Prof. T. Balasubramanian, Director and Prof. K. Kathiresan, Centre of Advanced Study in Marine Biology for their encouragement and the authorities of Annamalai University for the facilities provided.

References

Alongi, D. M., 1988. Bacterial productivity and microbial biomass in tropical mangrove sediments. *Microbial Ecology*, 15: 59-79.

Alongi, D. M., K. G. Boto, and F. Tirendi, 1989. Effect of exported mangrove litter on bacterial productivity and dissolved organic carbon fluxes in adjacent tropical nearshore sediments. *Marine Ecology Progress Series*, 56: 133-144.

Baumann, L., P. Baumann, M. Mandel and R. D. Allen, 1972. Taxonomy of aerobic marine eubacteria. *Journal of Bacteriology*, 110: 402-429.

Chandran, R. and K. Ramamoorthi, 1984. Hydrobiological studies in the gradient zone of the Vellar estuary. Part III. Heterotrophic bacteria. *Mahasagar*, 17: 197-120.

Cross, T., 1982. Actinomycetes: a continuing source of new metabolites. *Development in Industrial Microbiology*, 23: 1-18.

Das, S., P. S. Lyla and S. Ajmal Khan, 2004. Microbial diversity in mangrove and deep sea ecosystems. *Seshaiyana (ENVIS News Letter)*, 12 (1): 6-8.

Das, S., P. S. Lyla and S. Ajmal Khan, 2005. Variation in microbial population in a mangrove area of Parangipettai coast- Effect of Tsunami? *Research Journal of Chemistry and Environment*, 9 (3): 46-49.

Das, S., P. S. Lyla and S. Ajmal Khan, 2006. Mangrove ecosystems and the associated microorganisms. *Science India*, 9 (12): 15-17.

Holguin, G., P. Vazquez, and Y. Bashan, 2001. The role of sediment microorganisms in the productivity, conservation, and rehabilitation of mangrove ecosystems: an overview. *Biology and Fertility of Soils*, 33: 265-278.

Holt, J. G., N. R. Kreig, P. H. A. Sneath, J. T. Stanley and S. T. Williams, 1994. Bergey's manual of determinative Bacteriology. 9th Edition. The Williams and Wilkins, Baltimore, 787p.

Kannan, L., and K. Vasantha, 1986. Distribution of heterotrophic bacteria in Vellar estuary, East coast of India. *Indian Journal of Marine Sciences*, 15: 267-268.

Kathiresan, K. and B. L. Bingham, 2001. Biology of mangroves and mangrove ecosystems. *Advances in Marine Biology*, 40: 81-251.

Lakshmanaperumalsamy, P., D. Chandramohan and R. Natarajan, 1986. Seasonal variation of microbial population from sediments of Vellar estuary, south India. In: Proc. Deuxième Colloque International de Bactériologie marine, IFREMER, Actes de Colloques, 3: pp. 43-54.

Lechevalier, M.P. and H. Lechevalier, 1970. Chemical composition as a criterion in the classification of aerobic actinomycetes. *International Journal of Systematic Bacteriology*, 20: 435–443.

Odum, W. E. and E. J. Heald, 1975. Mangrove forests and aquatic productivity. *In*: Coupling of Land and Water Systems (ed. A. D. Hasler), Springer, Berlin, 129-136.

Oliver, J. D., 1982. Taxonomy scheme for the identification of marine bacteria. *Deep Sea Research*, 29: 795-798.

Prabu, S. K., B. Subramanian and A. Mahadevan, 1991. Occurrence and distribution of heterotrophic bacteria of Madras coast (Bay of Bengal). *Indian Journal of Marine Sciences*, 20: 130-133.

Prema, P., A. Palavesam and P. Balaji, 2005. Aerobic heterotrophic bacterial diversity in sediments of Rajakkamangalam estuary, south west coast of India. *Journal of Environmental Biology*, 26 (4): 729-734.

Purushothaman, A., 1999. Studies on bacteria ensemble in a developed vis-à-vis developing mangrove ecosystem. Ph.D thesis, Annamalai University, India, p. 124.

Ramaiah, N., C. Raghukumar, G. Sheelu and D. Chandramohan, 1996. Bacterial abundance, communities and heterotrophic activities in the coastal waters off Tamil Nadu. *Indian Journal of Marine Sciences*, 25: 234-239.

Ravel, J., M. J. Amoroso, R. R. Colwell and R. T. Hill, 1998. Mercury-resistant actinomycetes from the Chesapeake Bay. *FEMS Microbiology Letter*, 162: 177-184.

Sathiyamurthy, K. and A. Purushothaman, 1992. Distribution and composition of organic detritus in a tropical estuarine system. *Asian Marine Biology*, 9: 101-110.

Schneider, J. and G. Rheinheimer, 1988. Isolation methods. In: Methods in aquatic bacteriology, (ed. B. Austin), John Wiley and Sons Ltd., pp.73-94.

Shome, R., B. R. Shome, A. B. Mandal and A. K. Bandopadhyay, 1995. Bacterial flora in mangroves of Andaman. Part I: Isolation, identification and antibiogram studies. *Indian Journal of Marine Sciences*, 24: 97-98.

Simidu, U. and K. Aiso, 1962. Occurrence and distribution of heterotrophic bacteria in seawater from the Kamogewa Bay. *Bulletin of Japanese Society of Scientific Fisheries*, 28: 1133-1141.

Siva Kumar, K. 2001. Actinomycetes of an Indian mangrove (Pichavaram) environment: An inventory, Ph.D. thesis, Annamalai University, 91p.

Starr, M. P., H. Stolp, H. G. Turper, A. Balows and H. G. Schlegal, (eds.). 1981. The prokaryotes. A handbook on habitats, isolation and identification of bacteria. *Springer Verlag* (Vol 1&2): 2284p.

Vasantha, K. and L. Kannan, 1987. Distribution of heterotrophic bacteria in Kille backwaters, Porto Novo, Southeast coast of India. *Mahasagar*, 20: 35-41.

Walker, J. D and Colwell R. R, 1975. Factors affecting enumeration and isolation of actinomycetes from Chesapeake Bay and Southeastern Atlantic Ocean sediments. *Marine Biology*, 30: 193-201.

Chapter 4

A Study on the Physico-chemical Characteristics of Water in Vallur Village of Tiruvallur District, Tamil Nadu, India

M. Sangeetha, P. Lakshmi Devi, P. Indumathy, J. Jayanthi and M.G. Ragunathan

Introduction

Water is a gift to man by nature. But man's activities have made water as a source of potential danger to his health and very existence. The menace of water–borne diseases and epidemics still threatens the well being of population, particularly in under developed and developing countries. Thus, the quality as well as the quantity of clean water supply is of vital significance for the welfare of man kind (Anon, 1996).

Numerous anthropogenic activities like disposal of sewage and industrial water, recreational activities, excessive usage of fertilizers to lands and use of pesticides has threatened environmental health of both surface and ground water. In addition to this, due to the uneven distribution of fresh water on earth, due to developmental thrust of man and due to difference in the per capital supply and per capita availability of water, the water has become scarce natural resource and national wealth. Water pollution is now days considered not only in terms of public health but also in terms of its conservation, aesthetics and preservation of natural beauty and resources. Water pollution has however threatened to reduce the quantity available in ponds, lakes, rivers and reservoirs due to disposal of sewage, industrial water and due to other human activities (Trivedy and Chandrasekar, 1999).

Population explosion, industrialization, urbanization and developmental thrust of man have created these problems of water pollution. According to different surveys, 70 to 80 percent of Indian water sources are polluted and different enteric diseases affect million of the peoples every year.

Table 4.1: Physico–chemical Factors of Different Water Samples

Sl.No	Parameters	Normal/Maximum Acceptable Limits	Sample I	Sample II	Sample III	Sample IV	Sample V	Sample VI
1.	Appearance	Colourless	Turbid	Clear	Clear	Clear	Clear	Clear
2.	Odour	Unobjectionable	None	None	None	None	None	None
3.	Colour	Colourless	Colourless	Colourless	Colourless	Colourless	Colourless	Colourless
4.	Turbidity (NTU)	5–10	16.54	6.56	3.22	1.25	57.46	0.35
5.	Total dissolved solids (mg/L)	500–1500	115.8	334.5	3006.66	205.1	3820.1	7813.5
6.	Electrical Conductivity (Micro mho /cm)	1000–4770	158	483.9	4343.6	292	5522.5	11063.4
7.	pH	6.5–9.2	7.48	6.73	6.51	6.56	7.02	6.8
8.	Alkalinity (mg/L)	200–600	56	125.5	291.1	96.1	349.9	589.7
9.	Total hardness (mg/L)	300–600	51.4	113.7	1340.4	60.2	1342.9	2121
10.	Calcium (mg/L)	200–1000	11.4	25.2	335	15.5	342.9	522.5
11.	Magnesium (mg/L)	30	4.6	11.2	121.2	4.9	113.5	1945
12.	Sodium (mg/L)	150	14.4	55.2	381.9	42.5	639.2	1499.2
13.	Pottasium (mg/L)	12	4.	5.2	19.8	3.2	69.4	199
14.	Iron (mg/L)	1	1.42	0.648	0.514	0.313	0.637	0.36
15.	Maganese (mg/L)	0.05–0.17	Nil	Nil	Nil	Nil	Nil	Nil
16.	Free ammonia (mg/L)	3	0.568	0.57	0.328	0.255	0.226	0.616
17.	Nitrite (mg/L)	1	0.27	0.457	0.25	Nil	Nil	0.337
18.	Nitrate (mg/L)	45	4.4	4.4	2.7	4.9	3.2	59.5
19.	Chloride (mg/L)	250–1000	13.4	59.9	831.6	15.6	937.6	1929.4
20.	Fluoride (mg/L)	1–1.5	0.16	0.212	0.251	0.345	0.5	0.9
21.	Sulphate (mg/L)	200–450	44	12.6	783.8	21.9	1275	2478.8
22.	Phosphate (mg/L)	0.02–0.11	0.56	0.45	0.329	0.257	0.261	0.662
23.	Tidy's test (mg/L)	0.024	3.04	4.15	1.37	3.2	1.4	2.38
24.	Silica	–	19.59	15.22	57.49	26.5	37.42	34.52
25.	B.O.D (mg/L)	3.5	11.6	17.9	3.8	5	4.7	5.2

Today most of the water bodies of the world receive million of litres of sewage. Regular drinking water monitoring not only prevents disease and hazards but also checks the water resources from being further polluted (Meitei, *et al.*, 2004).Hence to check the water quality the present investigation has been undertaken.

Materials and Methods

The water sample is collected from six different sites. They are pond water I, pond water II, tap water I, tap water II, ground water I and ground water II. The water is collected in 2 litre clean plastic cans and brought to the laboratory and subjected to analysis. It is repeated for 10 times. The pond water I and pond water II located few kilometers from Vallur village. The tap water I and tap water II is coming from Chiruvakam filled in tank of Vallur village and are supplied to all over the village for drinking purpose, in the absence of the tap water the ground water is used for drinking purpose. The physico-chemical parameters as given in the Table 4.1 were done following APHA (1955). The data obtained were statistically analyzed and expressed as mean±SEM (Standard Error of Mean) using SPSS 7.5 Windows student's version software.

Results

The results of the different water samples were presented in the Table 4.1 against the standard values given by APHA (1985).

Discussion

The menace of water borne diseases and epidemics still threatens the well being of population, particularly in under developed and developing countries. Thus, the quality as well as the quantity of clean water supply is of vital significance for the welfare of mankind.

For example, the fabled Yamuna, on whose banks the Delhi City was born more than 2000 years ago, is a case study in the water management crisis confronting India. In Hindu, mythology, the Yamuna is considered to be a river that fell from heaven to earth. Today it is a foul portrait of crippled infra structure and yet worshipped. As the Yamuna enters the City (Delhi,) relatively clean from its 246 mile decent from top of Himalayan, the city's public water agency, the New Delhi Jal Board, extracts 229 million gallons every day from the river, as the river leaves the city it becomes the principle drain for New Delhi's waste. Residents pour 950 million gallons of sewage into the river each day (Sengupta, 2006).

The purpose of testing water samples is to interpret the data and take appropriate action. Apart from revealing the nature of water, the test results can throw light on the field condition, performance of treatment plants, presence of pollution, corrosion phenomenon etc. the utility of water quality data depends on how for it can be made use of for taking decisions on field problems.

In the present investigation when the results were analyzed it its clearly understood that the pond water I and II and tap water II is fit for drinking, tap water I and ground water I and II are highly polluted and unfit for drinking.

Referecnes

Anon.1996. A book on Training programme for new entrant junior water analysts conducted by RDT center, Tamil Nadu water supply and drainage board, Chennai.

A.P.H.A.. 1995. Standard methods for the examination of water and waste water, American Public Health Association.

Meitei, N.S., Bhargava, V. and Patil., P.M.2004. Water quality of Purna river in Purna town area of Maharastra state. *J. Aqua Biol*. 19 (2): 77-78.

Sengupta, S. 2006. In Delhi water means dry pipes, foul sludge. Deccan Chronicle, the daily newspaper. English edition, Chennai 3.10.2006: pp- 8

Trivedy, R.K. and Chandrasekar, T.R.1999. Sediment characteristics of fresh waterbodies of Mangalore, Karnataka. *J. Eco. Biol*. 11 (1): 59-64.

Chapter 5

Zooplankton Diversity of Goutami Godavari Estuary, Yanam, Pondicherry (U.T.)

K. Sree Latha and S. Rajalakshmi

Introduction

The distribution and variability of the principle nutrients in estuaries, largely determine the biomass and productivity of phytoplankton and zooplankton. They are natural inhabitants of waters and serve as the basis of food chain within the ecosystem. They are also involved in water pollution in a number of significant ways. In India the ecological aspects and the zooplankton studies have been known only in respect of some rivers, meager in the east coast (Srinivasan and Santhanam, 1991; Krishna Murthy and Subramanian, 1999; Chandra Mohan, 1963; Chandrashekar and Kodarkar, 1996, Narsimha Rao and Jaya Raju, 2001; Sita Rama Sharma, 1995; Prasad, 2003). The present paper deals with species diversity, composition of zooplankton in different seasons at two different stations.

The site selected for this study is Goutami Godavari estuary situated at southeast of Yanam (16° 42'and 16° 46' N and 82° 11' and 82° 19' E). The stretch of this estuary is 45 kms from the mouth of Goutami Godavari river extending between Kottapallem (near sea) and Kotipalle (at the head of the estuary). Two stations were selected, station A (Kanakalapeta which is 12 kms away from the sea) and Station B (Farampeta which is 7 kms away from sea), which is highly polluted zone.

The phytoplankton and zooplankton are always inversely proportional in an aquatic environment because the zooplankton feed on the phytoplankton. Zooplankton undertake journeys from bottom to surface at the approach of darkness, light intensity is considered the main factor and temperature, pressure, gravity and predators are also known to influence this phenomenon. Zooplankton serves as a link between primary production and tertiary production (forming major food source). They are also known to be indicators of the specific waters and the species, density of zooplankton is directly

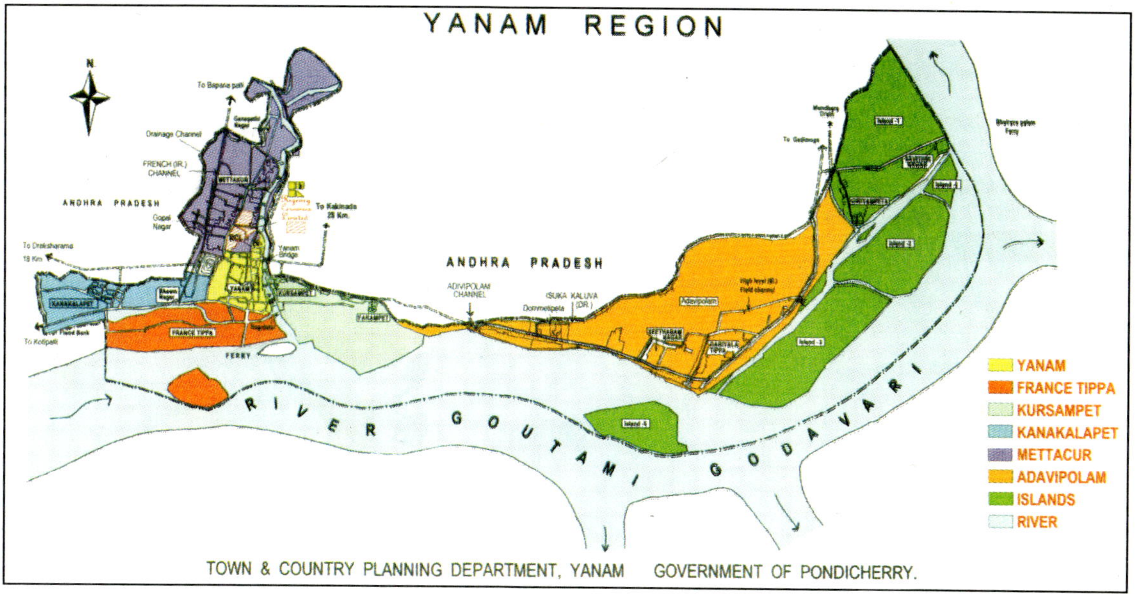

Figure 5.1: Yanam Region in Pondicherry

correlated with fishery potentiality (Sreelatha and Raja Lakshmi, 2005). The present investigation deals with the species diversity and composition of zooplankton of Goutami Godavari estuarine habitat.

Materials and Methods

Fortnight collections of zooplankton were made by using a half metre bottling nylon net number 21, mesh size 120µm from two stations (A and B) for a period of one year. The abiotic factors such as water temperature, transparency, pH, free carbon dioxide, dissolved oxygen, total alkalinity and chlorides were estimated using standard analytical methods (APHA, 1995). The samples were fixed with 4 per cent formalin, for quantitative estimation of the zooplankton was determined using Sedgwick rafter cell by taking 1 ml of diluted sample. Standard key and other literature was used for identification of different species.

Results and Discussion

Hydrographical Conditions

Estuarine environment exhibited wide range of physical and chemical factors in both time and space with marked variation from station to station. Physical factors such as Temperature, pH, Dissolved Oxygen, Free Carbon dioxide, Total Alkalinity, Sulphates, Chlorides and Total suspended solids (Table 5.1) control zooplankton diversity and density. During the present study, the transparency was relatively high at station A. The surface water temperature ranged from a minimum of 24.67ºC to 32.5ºC maximum, P^H from a minimum of 7.85 to 8.20 maximum, Total suspended solids ranged from170 to 1400. Salinity varied from total fresh water conditions124ppm to a normal estuarine salinity 250ppm. Chloride levels and Free Carbondioxide is high in polluted zone (Station B) due to human activities. Sulphates varied between 10.65 and 56.0. Dissolved oxygen between 0.5 ml/l and 14.6ml/l.

study than the past decades, this could be due to overall increase in the stress on zooplankton organisms, as a result of anthropogenic activities during the intervening years. Lack of stability in time to diversify and physiological stress imposed seems to be the main reason for both spatial as well as seasonal variation in diversity and richness of zooplankton.

Acknowledgements

The authors are grateful to Shri R. V. Gopala Rao, Prof and Head, Dept. of Zoology, and Principal, Dr. S. R. K. Govt. Arts College, Yanam., for providing laboratory facilities and valuable support.

References

APHA. 1995. Standard Methods for the examination of water and waste waters. 19th edition, New York., 874pp.

Chandra Mohan. 1963. Studies on zooplankton of Godavari estuary. Ph. D. Thesis, Andhra University, A. P. India.

Goswami, S. C. 1982. Distribution and Diversity of copepods in Mandovi Zuari estuarine system, Goa. *Indian J. Mar. Sci.*, 11 (4) 292-295.

Haridas P. Gopala Menon. P and Madhu pratap. M. 1980. Annual variations in zooplankton from a polluted coastal environment. *Mahasagar Bull. Nat. Inst. Oceanogr.*, 13 (3): 239-248.

Kelly, R., Jack. J. and Fant. 2000. A survey of the physio-chemical characteristics and zooplankton communities in an ephemeral Karst lake. *Aquatic Ecology*, 34 (1):77-89.

Krishna Murthy and Subramanian. 1999. Organisation of commercially supported meroplankton in palk Bay and Gulf of Mannar South east coast of India. *Indian J. Mar. Sci*, 28: 211-215.

Mishra and Saxena. 1992. *Aquatic Ecology*. Ashish Publishing House, New Delhi.

Mouny, P. and Dauvin. J. C. 2002. Environmental control of mesozooplankton community structure in seine Estuary (English Channel). *Oceanologica Acta, 25 (1):13-22.*

Narsimha Rao and Jaya Raju. 2001. Limnological investigations and Diversity of plankton in sewage fed fish culture at Nambur near Guntur A. P. *India. J. Aqua. Biol.* Vol. 16 (1 and 2):11-14.

Prasad N. V. 2003. Diversity and Richness of Zooplankton in Coringa Mangrove Ecosystem: Decadal changes. *J. Aqua. Biol.* 18 (2):41-46.

Sita Rama Sharma. 1995. Environmental impact studies on Teki-drain, Coringa river on plankton with reference to Sugar, distillery and paper mill effluents Ph. D. Thesis, Andhra University, A. P., India.

Sreelatha. K and Raja Lakshmi. S. 2005. Biodiversity of Phytoplankton diversity in Goutami Godavari Estuary atYanam, U.T. of Pondicherry. *J. Aqua. Biol.* 20 (2):45.

Chapter 6

Research on Bioactive Compounds Producing Marine Actinobacteria

K. Sivakumar, Maloy Kumar Sahu, J. Rajkumar, T. Thangaradjou and L. Kanna

Introduction

Actinobacteria have been looked upon as a potential source of bioactive compounds and the past experiences prove that these microbes are the richest sources of secondary metabolites. They hold a prominent position as targets in screening programs due to their diversity and their proven ability to produce novel metabolites and other molecules of pharmaceutical importance (Ellaiah *et al.*, 2004). Since the discovery of actinomycin (Lechevalier, 1982), actinobacteria have been researched and found to produce many commercially important bioactive compounds and antitumor agents in addition to enzymes of industrial interest (Tanaka and Omura, 1990; Umamaheswary *et al.*, 2005; Sivakumar *et al.*, 2006, Sahu *et al.*, 2007a,b,c; Murugan *et al.*, 2007). Approximately, two-thirds of the thousands of naturally occurring antibiotics have been isolated from these organisms (Takaizawa *et al.*, 1993). Of them, many have been obtained from *Streptomyces* (Goodfellow and O'Donnell, 1989) and these natural products have been an extraordinary source for lead structures in the development of new drugs.

So far, terrestrial soils were the most predominant and widely exploited source for actinobacteria. Investigations on marine actinobacteria are a few and inconclusive, though they are the important sources for the discovery of new bioactive compounds (Okami, 1984). In recent years, there has been a growing awareness of the potential value of marine sediments as sources of actinobacteria that produce useful bioactive metabolic products (Goodfellow and Haynes, 1984). Goodfellow and Hayens (1982) reviewed literature on isolation of actinobacteria from marine sediments and suggested that these sources may be valuable for the isolation of novel actinobacteria with the potential to yield useful new products. However, it has not yet been resolved whether the actinobacteria form part of the autochthonous marine microbial community of sediment samples originated from terrestrial habitats or they are simply carried to the sea in the form of resistant spores.

However, actinobacteria from marine samples have recently undergone screening for novel metabolites and there is evidence that actinobacteria usually make up only a small proportion of the bacterial flora of marine habitats with absolute numbers of actinobacteria much lower than the terrestrial habitats (Goodfellow and Williams, 1983). The consequence is that it would be more difficult to obtain large numbers of isolates from the marine samples for screening purposes. Some microbiologists have investigated the distribution and biological characteristics of the aquatic actinobacteria and their distribution was different from that of the soil actinobacteria (Kuster, 1976). Research on the biodiversity of marine actinobacteria is not only important as basic study but also necessary for their exploitation. Still now, there are no comprehensive data on the bioactive compounds producing marine actinobacteria in the global scenario and hence the present attempt to present a review of available literature on bioactive compounds producing marine actinobacteria.

Bioactive Compounds from Marine Actinobacteria

Knowledge of the antibiotic producing marine actinobacteria, is almost entirely the result of the pioneering work of researchers at the Institute of Microbial Chemistry in Tokyo. Beginning in the 1970s, Okami and co-waorkers did innovative work in isolating bacteria from both the shallow and deep-water sediments from the coasts of Japan. Their work, summarized in several reviews, provides the foundation for the development of this field (Okami, 1986; 1988). In 1972, Institute of Microbial Chemistry group isolated 136 strains of actinobacteria from 37 different samples collected from three inlets in Sagami Bay and tested for their anti-microbial and anti-cancer activities. Among 136 strains tested, 37 strains (27 per cent) and 25 strains showed very high anti-microbial and anti-cancer activities respectively. The results suggested that the marine actinobacteria are very useful screening sources of anti-microbial and anti-cancer compounds (Okazaki and Okami, 1972). In 1975, the same group (Okazaki *et al.*, 1975) reported an antibiotic-producing actinobacteria, *Chainia purpurogena* SS-228 isolated from mud samples from Sagami Bay. The organism was found to produce the benzanthraaquinone antibiotic which showed selective inhibition of gram-positive anti-tumor bacteria, with MIC values ranging between 1-2 mg/mL and was an active agent *in vivo* against Ehrlich carcinoma in mice, showing significant lifespan extension at low doses. In addition, quinnone deactivated the dopamine-b hydroxylase (producing 65 per cent inhibition at 0.1 mg/mL). Most interestingly, the production of quinnone was only observed in selected seawater media containing the unique Japanese seaweed product "Kobu Cha" produced in Japan from the brown seaweed *Laminaria*. This paper was among the first to show that bacteria of order Actinomycetales could be isolated from marine environment. Another significant aspect of the *Chainia purpurogena* antibiotic study was the specificity of production of antibiotic only under selected, marine-specific nutrient conditions. This observation clearly showed that marine microorganisms have nutrient adaptations which relate to their natural nutrients in marine habitats (Jensen *et al.*, 1991). Skarbek and Brady (1975) isolated chromomycin-A3 antibiotic from a new subspecies of *Streptomyces* that showed activity against gram-positive bacteria and mycobacteria. A new antibiotic, aplasmomycin, which inhibits the growth of gram-positive bacteria including myobacteria *in vitro*, and plasmodia *in vivo* was obtained from a strain of *Streptomyces griseus* isolated from shallow sea sediment in Sagami Bay. The antibiotic aplasmomycin forms colourless needle-like crystals and has a molecular formula of C41H60O14Na (Okami *et al.*, 1976). The istaycines A and B are other examples of antibiotics isolated by the Tokyo group from *S. tenjimariensis* from a shallow-water mud sample from Sagami Bay, Japan. (Okami, 1979). The compounds showed strong *in vitro* antibiotic activity against both gram-negative and gram-positive bacteria, including some strains which are known to be resistant to the aplasmomycins A-C antibiotics. Istaycines A and B showed MIC values ranging between 0.10 and 3.0 mg/mL against various species and isolates of *Staphylococcus, Bacillus, Corynebacterium* and *Escherichia*. The compounds were much less active against *Pseudomonas, Klebsiella* and *Serratia* species. The istaycines are related to the fortimicines, sporaricins and aminoglycoside antibiotics produced by terrestrial actinobacteria. They appear more therapeutically important, because of their activity against typical aminoglycoside-resistant pathogens.

Among the more unusual compounds isolated from marine actinobacteria, aplasmomycins A-C were also reported by the Tokyo group. These compounds were produced by a marine actinobacteria identified as *Streptomyces griseus* SS-20, isolated from shallow mud samples also from Sagami Bay (Okami *et al.*, 1976; Hotta *et al.*, 1980). Here too, this organism alone produced the aplasmomycins when cultured with Kobu Cha-containing media or when grown under conditions (27°C and very low nutrients) which relate to the natural environment of Sagami Bay. The aplasmomycines are antibiotics which inhibit gram-positive bacteria in vitro with MIC values ranging from 0.8 to 3.0 mg/mL aganist *Staphylococcus aureus, Bacillus subtilis, B. anthracis* and *Corynebacterium smegmatis*. More importantly, aplasmomycin is an effective antimalarial agent in vitro, resulting in cures of Plasmodium bergheri-infected mice. It was on the basis of this potent antiplsmodium activity, aplasmomycin received its name. The structure eludication of aplasmomycin A was performed by X-ray crystallographic methods. To the surprise of the Tokyo group, the molecule contained an unexpected boron atom in the centre of the complex (Nakamura *et al.*, 1977). Though quite remarkable structures, these compounds are precedent by the boromycines, similar boron-containing ionophores produced by a terrestrial actinobacteria. An aplasmomycin C-producing actinobacteria was isolated from sandy sediment samples in California. Unexpectedly, the structure of aplsmomycin C was found to be even more complex than originally observed. By NMR and X-ray methods, aplasmomycin C was found to exist in several chromatographical-stable ring conformers (Stout *et al.*, 1991).

Laksmanaperumalsamy *et al.* (1978) isolated 518 streptomycetes strains from the sediments of estuarine, backwater, marine, freshwater and mangrove environment of Porto Novo, India using Grein and Meyer's agar, Kuster's agar and Glucose Asparaginase agar. These isolates were checked for both antibacterial and antifungal activities by cross assay method against *B. circulans, S. aureus, E. coli, P. aeruginosa, S. cerevisiae* and *F. oxysporum*. It was found that, 27.03 per cent of the strains was found to elaborate one or more types of antibiotics and 59.27 per cent was active against *B. circulans*, 47.01 per cent against *S. aureus*, 30 per cent against *E. coli*, 53.59 per cent against *S. cerevisiae* and 39.3 per cent against *F. oxysporum*. Majority of the isolates (46.43 per cent) showed combined antibacterial and antifungal activity and 25 per cent showed only antibacterial activity. Unidentified estuarine actinobacteria from California were found to produce two metabolites of a unique mixed biosysnthesis origin. Marinone and debromomarinone (Nakamura *et al.*, 1977; Hotta *et al.*, 1980) are antibacterial metabolites possessing common naphthoquinone, but rare sesquiterpenoid structural compounds. In addition, marinone (Hotta *et al.*, 1980) possesses a bromine substituent in the dihydroxybenzene ring, a position typical for bromination in marine metabolites. Both molecules show antibacterial activities, in the range of 1-2 mg/ml, against gram-positive bacteria of the genera *Bacillus* and *Staphylococcus* (Pathirana *et al.*, 1993). The Tokyo group has also reported manipulations of the expression of antibiotic synthesis in some marine actinobacteria. When protoplasts of the istamicin-producing strain *S. tenjimariensis* SS-939 were fused with the aplasmomycin-producing *S.giseus* SS-20, a new clone, SK2-52, was obtained which produced a new indolizine antibiotic, indolizomycin (Okami, 1988; Gomi, 1984; Yamashita, 1985).

Pisano *et al.* (1986) screened 120 strains of marine actinobacteria from sediments of the Sandy Hook Bay, New Jersey by pretreatment methods. Among the 120 strains, 19 strains displayed significant antimicrobial activity. Most of the activity was directed against gram-positive bacteria, but inhibition of gram-negative species and yeast was also evident. The same group Pisano *et al.* (1989) isolated bioactive actinobacteria from marine sediments using starch-casein agar, supplemented with rifampicin. *Streptomyces* and *Micromonospora* represented the major genera identified. The isolated strains showed antimicrobial activity against both gram-positive and gram-negative bacteria. The results suggested that rifampicin, when added to starch-casein agar would be effective for the isolation of bioactive actinobacteria from marine sediments. Takahashi *et al.* (1989a; 1989b) isolated a structurally-novel alkaloid, altemicidin, which is produced by a marine strain of *Streptomyces sioyaensis* SA-1758. This exceptional alkaloid was detected by screening cultures for toxicity against the common brine shrimp *Artemia salina*. Earlier reports have shown that Artemia toxicity translates amazingly well to antitumor activity (Meyer *et al.*, 1982). Indeed, altermicidin showed potent antitumor activity in vitro against L1210 murine leukemia and IMC carcinoma cell lines with IC_{50} values of 0.84 and 0.82 mg/mL, respectively. The new compound showed weak antibacterial activity, but was relatively toxic in mice (LD_{50} = 0.3 mg/kg iv), propablbly limiting its chemptherapeutic uses. Altemicidin is an exceptionally novel sulfer- and nitrogen-containing microbial metabolitie. While its biosynthesis has not been defined, altemicidin possesses a monoterpene carbon skeleton, indicating that it may be produced via the mevalonate pathway. Fenical *et al.* (1991) studied the actinobacteria (CNB-032) isolated from the shallow sediments of Bodega Bay, California. On the basis of its cell wall constituents and whole cell sugar composition, this organisms keyd to a member of the suprageneric group Maduramycetes. Fermentation of this organism resulted in the production of maduralide, a new member of a rare class of macrolide. Although the biological properties of this macrolide are not yet known, the compound is a member of a rare 24-membered ring lactone group represented previously only by rectilavendomycin (Osmura, 1984). The compound was not crystalline and hence X-ray methods could not be applied to the structure determination. The structure assignment, yielding the molecular structure without stereo-chemical details, was determined by a combined spectral method. Balagurunathan *et al.* (1989) studied the antagonistic actinobacteria isolated from the littoral sediments of Parangipettai. Among the 51 strains, only 11 strains showed good antibiotic activity and they were identified as *Streptomyces* spp. and *Nocardia* spp. Furthermore, they (Balagurunathan *et al.*, 1989) isolated Y-lactone type of antibiotic from *Streptomyces griseobrunneus*. This antibiotic was tested against fish pathogens *viz.* species of *Vibrio, Aeromonas, Pseudomonas, Bacillus* and *Fusarium* and it inhibited all these pathogenic organisms with inhibition zones ranging from 10-30 mm. The *Vibrio* sp. and *Pseuomonas* sp. were more sensitive than the other bacterial species tested.

Pathirana *et al.* (1992) isolated *Streptomyces* sp. from Bodega Bay, California, which produced some possibly useful antibacterial compounds. Purification of the extract of the whole fermentation broth yielded four new compounds of the phenazine class. These compounds were phenazine esters of the rare sugars, L-quinovose, at the sugar 2' and 3' positions. Although the alpha and beta anomers in each series were readily isolated by silica HPLC, they were found to be interconvertents. The new metabolites showed antibacterial activities against a range, with MIC values in the 1-4 mg/mL. They were not appreciably cytotoxic against human tumor cells *in vitro.* In the same group (Pathirana *et al.*, 1991) reported actinobacteria strain from the coral reef carbonate sediment of Bahamas which produced C_{13} butanolide. The simple butanolide was closely related to a series of signal compounds previously known to regulate secondary metabolite synthesis by these microorganisms. As part of the structural ekudication process for butanolide, an enantioselective synthesis was performed. Vanajakumar *et al.* (1991) screened 386 strains of actinobacteria from five marine molluscs *viz., Crassostrea madrasensis,*

Meretrix casta, Anadara rhombea, Telescopium telescopium and *Bullia vittata* from the shell surface, mantle and gut of all the five molluscs. Mantle tissue harboured the most antagonistic strains (84 per cent), followed by gut (77 per cent) and shell surface (69 per cent). When these strains were screened for the production of antibiotics by cross-streak method, 290 strains (75 per cent) exhibited antagonistic properties. Combined antibacterial and fungal properties were found in 46.4 per cent of the antagonistic actinobacteria. Cultures exhibiting only antifungal properties were found in 2.4 per cent of the actinobacteria while antibacterial activities were seen in 42 per cent. In the colour-series tests, 99 (84 per cent) of the grey-series, 63 (82 per cent) of the 77 yellow series, 11 (73 per cent) of the 15 red series and 117 (66 per cent) of the 176 white series showed antagonistic properties. Tapiolas *et al.* (1991) reported streptomycete from the surface of a gorgonian coral (*Pacifigorgia* sp.) from the Gulf of Califonia, Mexico. This isolate (strain PG-19), when grown in marine media, produced a series of undescribed metabolites possessing cytotoxic and antibiotic properties. These compounds possess antibiotic properties similar to their parent molecules. In addition, this organism yielded two totally new cytotoxic agents, octalactins A and B, belonging to a new structure class. The compounds are C_{19} ketones possessing rare 8-membered ring lactone functionalities. Octalactin A possesses potent *in vitro* cytotoxicity against B16-F10 murine melanoma (IC_{50} = 7.2 x 10^{-3} mg/mL) and HCT-116 human carcinoma (IC_{50} = 0.5 mg/mL) cell lines. Octalactine B is devoid of cytotoxic activity, leading to the conclusion that the eposy keton functionality is essential for cytotoxicity. The structures of these compounds were finalized by X-ray studies on octalactine A, but the absolute stereochemistry was not defined.

Jellyfish associated *Streptomyces* sp. was reported by Trischim *et al.* (1993) which showed significant antibiotic activity. Investigation of the source of this activity led to the isolation of two new bicyclic peptides, salnamides A and B, which possess novel depsipeptide backdrops. The structure of salinamides A was derived first by spectral methods and by hydrolysis to yield several chiral amino acids. Subsequently, the crystalline isomer, salinamide B, was isolated and its structure was determined by X-ray methods. Conversion salnamides A and B with HCL in methanol further linked these molecules confirming both their relative and absolute stereochemistries. The salinimides are antibiotics of moderate potency with selective activity against gram-positive bacteria. The phenyl ether ring illustrates restricted rotation with all four aromatic portions shown at distinct NMR chemical shifts. This property and its selective antibiotic activity have opened the question of its relatedness to the vancomycin antibiotics which act by binding peptide cell wall precursors. Pathirana *et al.* (1993) isolated an unidentified estuarine actinomycete, from California which produced two metabolites of a unique mixed biosynthetic origin. Marinone and debromomarinon are antibacterial metabolites possessing common naphthoquinone, but rare sesquiterpenoid structural compounds. In addition, marinone possesses a bromine typical for bromination in marine metabolities. Both molecules show antibacterial activities, in the range of 1-2 mg/mL against Gram-positive bacteria of the genera *Bacillus* and *Staphylococcus*.

An actinomycete strain of the *Streptomyces* genus isolated by Bernan *et al.* (1994) from an intertidal sediment sample collected from the Key West, Florida, produced a series of new antibiotics called the bioxalomycins. The new antibiotics showed excellent antimicrobial activity against *Staphylococcus* and *Enterococcus* spp. Fermentation conditions for production of bioxalomycin alpha differed substantially from those required for production of a related compound, naphthyridinomycin, by the reference culture *Streptomyces slusitanus* NRRL 8034. Mathew *et al.* (1994) isolated *Streptomyces* spp. from *Villorita cyprinoids* and checked their antagonistic activity against *V. anguillarum, S. aureus, C. albicans, A. niger* and *S. cerevisiae*. Sitachitta *et al.* (1996) reported four new a-pyrone-containing metabolites, wailupemycins A-C (l-3) and 3-epi-Sdeoxyenterocin from a *Strepromyces* sp. collected from shallow water marine sediments. The structures of the new compounds were determined through the interpretation of spectral data. Compound 1 exhibited antimicrobial activity towards the Gram-negative bacterium *Escherichia coli,* while compound 4 inhibited the growth of *Staphylococcus aureus.* Ellaiah *et al.* (1996) isolated 28 strains of actinobacteria from marine sediments off Madras coast. Among them, three isolates belonging *Streptomyces* exhibited very good broad spectrum antibacterial activity. A novel bioactive depsipeptide, thiocoraline produced by *Micromonospora* was isolated from the mycelial cake (Romero *et al.,* 1997). The compound showed a potent cytotoxic activity against P-388, A-549 and MEL-28 cell lines and also a strong antimicrobial activity against gram-positive microorganisms. This compound binds to supercoiled DNA and inhibits RNA synthesis.

Pusecker *et al.* (1997) isolated unidentified marine *Streptomyces* sp. which produce novel 5,10-dihydrophencomycin methyl ester (4) and the known microbial metabolites (2-hydroxyphenyl)-acetamide (1), menaquinone MK9 (II, III, VIII, IX-H8) (2), and phencomycin (3a). Compound 4 showed weak antibiotic activity *against Escherichia coli* and *Bacillus subtilis.* Jrumg *et al.* (1997) reported Actinoflavoside, a novel glycoside composed of a rare 2, 3, 6-tridcoxy-3-amino-ribo- pyranoside (ristosamino) aminosugar and an aglycone reminiscent of the plant flavonoids from the culture broth of a marine *Streptomyces* sp. Similarly, Biabani *et al.* (1997) repoted Delta-Indomycinone, a new member of pluramycin class of antibiotics isolated from marine *Streptomyces* sp. The same group repoted 2-[Methyl-(3-phenylpropionyl)amino]-benzoic acid new antimicrobial compounds from a culture of marine *Streptomyces* sp. strain B7747. Analogous compounds have potential importance as phytotoxic substances. The Minimum Inhibitory Concentration (MIC) against *Chlorella vulgaris, Chlorella sorokiniana, Chlorella salina* and *Scenedesmus subspicatus* ranged from 20 to 107 micrograms/ml. All anthranilamides were inactive against *Staphylococcus aureus, Escherichia coli* and *Mucor miehei.* Arenaric acid (1a), a new pentacyclic polyether related to the antibiotics K-41A and oxolonomycin, was isolated as its sodium salt (1b) from the culture broth of an estuarine bacterial isolate of the genus *Streptomyces* (Cheng *et al.,* 1999). In 1999, Moore and co-workers reported antiinflammatory agents, salinamides A and B from the marine isolate *Streptomyces* sp. CNB-091. Total structures of the compounds were established using a combination of spectral and chemical techniques. Revised structures were presented for the bicyclic depsipeptides salinamides A and B on the basis of the analysis of the dansylated salinamide A hydrolysate by chiral capillary electrophoresis. The fermentation yield of salinamide D, which contains a D-valine residue in place of the D-isoleucine moiety in salinamide A, can be dramatically increased 30-fold by supplementing the growth media with L-valine. Salinamides C and E are monocyclic depsipeptides that are likely methylated byproducts of salinamide A biosynthetic intermediates. Holyrines A and B, two new members of the indolocarbazole class of alkaloids were produced by marine actinobacteria collected from the North Atlantic Ocean (David *et al.,* 1999). Holyrines A and B are possible intermediates in staurosporine biosynthesis.

Fernandez-Chimeno *et al.* (2000) reported a novel bioactive macrolide, IB-96212 from the fermentation broth of a marine *Micromonospora* sp. The macrolide showed a very strong cytotoxic activity against P-388, and lower but significant activity against A-549, HT-29 and MEL-28 cell lines. IB-96212, another novel cytotoxic macrolide was isolated from marine *Micromonospora* by Canedo *et al.* (2000). Structure of this compound consists of a new aglycone which possesses a 26-membered macrolide ring system and one deoxy sugar identified as L-rhodinose. This structure represents the first reported spiroketal macrolide natural product related to other macrolides, such as oligomycins, dunaimycins, citovaricins, rutamycins and ossamycins. Ingo *et al.* (2000a) isolated a marine actinobacteria which produced Neomarinone, a novel metabolite possessing a new sesquiterpene-

and polyketide-derived carbon skeleton, and several derivatives, of the marinone class of naphthoquinone aibiotics. Neomarinone and several of the marinone derivatives were shown to be moderately cytotoxic towards human cancer cells *in vitro* testing. Furthermore, Ingo *et al.* (2000b) isolated 439 actinobacterial strains from marine organisms, collected from Xiamen Island, China. Among them, 20.6 per cent of actinobacteria cultures displayed cytotoxic activity on P388 cells at dilutions at and below 1:320; 18.6 per cent on KB cells and 2.96 per cent of actinobacterial cultures displayed inducing activity. Among all actinobacteria isolated, the genus *Micromonospora* had the highest positive rate of inducing activity. However, most antimicrobial activities were found in the genus *Streptomyces*. These results indicate that actinobacteria associated with marine organisms could be a promising source for antitumor and antimicrobial bioactive substances. Hertwecka and Moore (2000) characterized for the first time, a plant-like biosynthesis of benzoic acid from the marine actinobacteria, *Streptomyces maritimus*. Feeding experiments with 2H- and 13C-labeled intermediates revealed that phenylalanine is metabolized to benzoyl-coenzyme A (CoA) by means of a phenylalanine ammonia lyase and subsequent b-oxidation of cinnamoyl-CoA. Benzoyl-CoA serves as the rare starter unit of a type II polyketide synthase producing the structurally novel bacteriostatic polyketides enterocin and the wailupemycins. The results from the feeding study were consistent with the proposed biosynthetic model deduced from the cloned enterocin biosynthesis gene cluster. Mukku *et al.* (2000) isolated two marine streptomycetes which produced four new butenolides, namely 4, 10-dihydroxy-10-methyl-dodec-2-en-1,4-olide (1), two diastereomeric 4,11-dihydroxy-10-methyl-dodec-2-en-1,4-olides (2/3), and 4-hydroxy-10-methyl-11-oxo-dodec-2-en-1,4-olide (4). The structures were identified by interpretation of the 2D NMR and mass spectral data.

A protease-producing, crude oil degrading marine *Nocardiopsis* sp. was reported by Dixit and Pant (2000). The *Nocardiopsis* produces two extracellular proteases, both of which are alkaline serine endopeptidases. Protease I was purified to homogeneity by chromatography on CM-Sephadex at pH 5.0 and pH 9.0. Protease II was purified using DEAE-cellulose, Sephadex G-50, phenyl-Sepharose and hydroxyapatite chromatography. Protease I and II had almost similar M(r) of 21 kDa (Protease I) and 23 kDa (Protease II), pI of 8.3 and 7.0 respectively with pH and temperature optima for activity between 10.0 and 11.0 and about 60 degrees C. Specific activities were 152 and 14 U/mg respectively on casein. However, Protease I was antigenically unrelated to Protease II. Both proteases were endopeptidases and required extended substrate binding for catalysis. Both proteases had collagenolytic and fibrinolytic activity but only Protease I had elastinolytic activity. The proteases were chymotrypsin-like with respect to their amino acid compositions and N-terminal sequences. Hernandez *et al.* (2000) reported N-acetyl-gamma-hydroxyvaline lactone, an unusual amino acid derivative from a streptomycete isolated from marine sediments of north coastline of Sao Paulo State, Brazil. *Microbacterium* sp., isolated from the marine sponge *Halichondria panicea*, produced four unusual cell-associated glycoglycerolipids and one diphosphatidylglycerol when grown on marine broth and on artificial seawater media (Wicke *et al.*, 2000). The lipids were isolated by chromatography on silica columns and their structures elucidated using a combination of multidimensional NMR and MS techniques. The main compound was 1-O-acyl-3-[alpha-glucopyranosyl-(1-3)-(6-O-acyl-alpha-mannopyranosyl)]glycerol (GGL.2) with 14-methyl-hexadecanoic acid and 12-methyl-tetradecanoic acid positioned at C-6 of the mannose unit and at the glycerol moiety. Glycolipid production was correlated with growth which reached a maximum value of 200 mg/L when grown on artificial seawater medium with 20 g/L glucose. The main compound decreased the surface tension of water down to 33 mN/m and the interfacial tension of the water/n-hexadecane system down to 5 mN/m. In addition to this good surface-active behaviour, the main glycoglycerolipid showed antitumor activities.

Hernan *et al.* (2000) reported two new indolocarbazole alkaloids, 4'-N-methyl-5'-hydroxystaurosporine (2) and 5'-hydroxystaurosporine (3), together with the known staurosporine (1) from the culture broth of a marine *Micromonospora* sp. The compounds showed cytotoxic activity against various tumor cell lines. A marine actinobacteria from a sample of beach sand collected near Lorne on the southwest coast of Victoria, Australia, yielded two new aromatic amides, lorneamide A (1) and lorneamide B (2) (Capon *et al.*, 2000), belonging to a novel class of tri-alkyl-substituted benzenes and their structures were determined by spectroscopic methods. Partil *et al.* (2001a) reported 133 strains of actinobacteria from 129 marine samples collected from various stations along the Tuticorin coast. Of the 104 strains of actinobacteria screened for the inhibitory activity against bacterial pathogens associated with fish diseases *viz., Aeromonas hydrophila, Aeromonas sobria* and *Edwardsiella tarda*, 77 isolates were inhibitory to at least one of the pathogens. The highest incidence of inhibitory isolates was noticed in the sediment samples and all the isolates of antagonistic marine actinobacteria were *Streptomytces* spp. Balagurunathan and Subramanian (2001) isolated 51 strains of *Streptomyces* from the littoral sediments of Parangipettai coastal waters. Out of these, only 8 strains showed very promising antibiotic activity against bacteria and fungi. These strains exhibited higher activity against gram-positive bacteria than the gram-negative bacteria. The strains also showed positive chitinase, protease and cellulase activities. Patil *et al.* (2001b) isolated 20 actinobacteria strains from the water and sediment samples of mangrove area of Tuticorin. The average actinobacteria load in the water and sediments was $4.79x10^4$ CFU/ml and $5.03x10^4$ CFU/g respectively. The strains were checked for their antagonistic activity against seven shrimp bacterial pathogens. Among them, 83 per cent showed good antagonistic activity against all the tested pathogens.

Schumacher *et al.* (2001) isolated two new indole nucleosides kahakamides A (1) and B (2) from the actinobacteria *Nocardiopsis dassonvillei*, obtained from a shallow water sediment sample collected on the island of Kauai, Hawaii. Compounds 1 and 2 are related to the neosidomycin antibiotics, a group of rare indole-*N*-glycosides. Compound 1 exhibited antimicrobial activity towards the Gram-positive bacterium *Bacillus subtilis*. Six novel metabolites containing a lactone moiety as a common structural feature, along with the previously described (-)-blastmycinolactol was isolated from the culture broth of sediment derived marine *Streptomyces* sp. (Cho *et al.*, 2001). On the basis of the results of combined spectroscopic analysis, structures of the new compounds were determined as butenolides and 3-hydroxy-gamma-butyrolactones. In shallow sea sediments from the region of Livingston Island of Antarctica, Ivanova *et al.* (2001) isolated *Streptomyces* sp. which produced compounds identical to phthalic acid diethyl ester (C12H14O4, MW. 222); 1, 3-bis (3-phenoxyphenoxy)benzene (C30H22O4, MW.446); hexanedioic acid dioctyl ester (C22H42O4, MW.370) and the new substance 2-amino- 9, 13–dimethyl heptadecanoic acid (C19H39NO2, MW.313). These compounds represent diverse classes of chemical structures and provide with evidence for the untapped biosynthetic potential of marine bacteria from Antarctica. Woo *et al.* (2002) found a novel antifungal protein (SAP) from the culture supernatant of a marine *Streptomyces* sp. This protein was characterized by chemical, biochemical, and biological analyses. The molecular weight of SAP was about 160 kDa. The SAP was composed of three heterologous protein subunits of 41.7 kDa (SAP1), 21.7 kDa (SAP2), and 18.7 kDa (SAP3) at a molar ratio of 1:1:5 (or 1:1:6). N-terminal amino acid sequence analysis and a homology search revealed that SAP1, SAP2, and SAP3 exhibit 64.3, 68.4, and 86.7 per cent similarity to three *Streptomyces* coelicolor polypeptides, puromycin resistance protein (Pur8), a conserved hypothetical protein, and bacterioferritin, respectively. The MIC of purified SAP against *Pythium porphyrae* was determined to be 1.6 microg/disk, whereas no inhibitory effect was observed at concentrations up to 100 microg/disk against most of the fungal and bacterial strains tested; the only exception was relatively strong antifungal activity against *Pythium ultimum* (MIC, 6.3 microg/disk). *In vitro* and *in vivo* toxicity tests demonstrated that SAP showed no toxicity against *Porphyra yezoensis* cells, human normal dermal fibroblasts, and mice at doses up to 700 microg/ml (for 24 h), 250 microg/ml (for 12 h), and 75 mg/kg (for 35 days), respectively. SAP was labile when it was subjected to a heated-air drying treatment, which is a great advantage in food production procedures. These studies indicate that *Streptomyces* sp. strain might be useful as a gene source for safe transgenic *Porphyra* breeding, to tolertate *Pythium* infection.

Dhevendaran and Anithakumari (2002) reported 250 species of streptomycetes from the gut of the fish, *Therapon jarbua* and shellfish, *Villorita cyprinoids* and sediments of Veli lake. Among them, 16 isolates were randomly selected to check their antagonistic activity against *Vibrio anguillarum, V. alginolyticus, V. parahaemolyticus, S. aureus, A. niger, C. albicans* and *S. cerevisiae.* Of these, 12 isolates inhibited the growth of *V. anguillarum*, whereas 10 isolates showed inhibition to *S. aureus*. None of the isolates inhibited the growth of *V. alginolyticus, V. parahaemolyticus, A. niger, C. albicans* and *S. cerevisiae.* Maskey *et al.* (2002) reported a new antibiotic with novel structure designated as parimycin from the culture broth of marine *Streptomyces* sp. The structure of the new antibiotic was determined by spectroscopic methods and by comparison of the NMR data with those of the structurally related gamma-indomycinone. A new macrolide antibiotic designated as chalcomycin B (1b) was isolated from the culture broth of a marine *Streptomyces* sp. together with chalcomycin (1a) as active principles (Asolkar *et al.*, 2002). The structure of the new antibiotic was determined by EI and ESI MS, 1H, 13C and 2D NMR spectroscopy and by comparison of the NMR data with those of chalcomycin. Yamada *et al.* (2002) isolated *Streptomyces hygroscopicus* from the marine fish *Halichoeres bleekeri*. The isolated strain produced Halichoblelide, a novel macrolide with potent cytotoxicity against tumour cells. The absolute stereostructure of the compound was elucidated on the basis of spectroscopic analyses using 1D and 2D NMR techniques and some chemical transformations. Salinosporamide A, a highly cytotoxic proteasome inhibitor derived from a novel marine *Salinospora* sp. was reported by Feling *et al.* (2003). Shin *et al.* (2003) screened a new cyclic tetrapeptide (1) from *Nocardiopsis* sp. isolated from the Pacific deep-sea sediment. Structure of this compound was determined as cyclo-(L-isoleucyl-L-prolyl-L-leucyl-L-prolyl) on the basis of combined chemical and spectral methods. Three novel antibiotics designated as chandrananimycin A (3c), B (3d) and C (4) were isolated from marine *Actinomadura* sp. (Maskey *et al.*, 2003). Structures of these new antibiotics were determined by detailed interpretation of mass, 1D and 2D NMR spectra. Similarly Itoh *et al.* (2003a, b) reported a novel anthracycline, komodoquinone A (1) and its aglycone, komodoquinone B (2) isolated from the sediment derived marine *Streptomyces* sp. Komodoquinone A (1) is a unique anthracycline, in which a new amino sugar is connected to the D-ring of the anthracyclinone skeleton, and was found to induce neuronal cell differentiation in the neuroblastoma cell line and Neuro 2A. Two new anthracycline antibiotics designated as himalomycin A (2c) and B (2d) were isolated from a marine *Streptomyces* sp. (Maskey *et al.*, 2003). The structure of the new antibiotics was determined by comparison of the NMR data with those of fridamycin D (2b) and by detailed interpretation of mass, 1D and 2D NMR spectra. Mathew and Philip (2003) isolated six strains of actinobacteria from the sediments of the Arabian sea, India. The isolated strains were tested for their antibiotic production using 14 different fermentation media. The strains which produced antibiotics in all these fermentation media were collected and checked for their nature of inhibition zones against 22 bacterial pathogens (*Vibrio* spp.). Of the 14 fermentation media used, only M_{13} and M_{14} media were found to be best in terms of their support to antibiotic production. Xiong *et al.* (2004; 2005) isolated a total of 331 *Streptomyces* sp. from seawater and sea sediments of Beidiahe and Dagang, China. The isolates were screened for their insecticidal qualities. The screening was carried out using bioassay of brine shrimp and the insect pest *Helicoverpa armigera*. Of the 40 isolates, one isolate, *Streptomyces avermitilis* was found to have strong insecticidal activity against both brine shrimp and *Helicoverpa armigera*. This study indicated that the marine *Streptomyces* sp. could be an important source of insecticidal antibiotics and the improved anti-brine shrimp bioassay is suitable for primary screening.

A gram-positive bacterial strain containing L-ornithine as the diagnostic diamino acid was isolated from a sea-water-sample from the East Sea, Korea (Yi *et al.*, 2004). A phylogenetic analysis based on 16S rRNA gene sequences showed that strain JC1078T represents a phyletic line within the suborder Micrococcineae of the order Actinomycetales, adjacent to the genus *Ornithinimicrobium*. The highest sequence similarity values to the isolate were observed against *Ornithinimicrobium humiphilum* (94.3 per cent) and *Kytococcus sedentarius* (94.1 per cent). The strain was strictly aerobic and moderately halophilic with optimal growth at 2-3 per cent (w/v) NaCl. Cells were non-motile, non-sporulating and coccoid-shaped. The cell wall contained L-ornithine, glutamic acid, alanine, glycine and serine. The major menaquinone was MK-8(H4). The predominant cellular fatty acids were of the iso- and anteiso-methyl-branched types. The polar lipids were phosphatidylglycerol, diphosphatidylglycerol, phosphatidylinositol and an unknown glycolipid. The acyl type of the glycan chain of peptidoglycan was acetyl. The DNA G + C content was 72 mol per cent. The combination of physiological, biochemical and chemotaxonomical data clearly separated the marine isolate from other members of the suborder Micrococcineae. On the basis of polyphasic evidence, they classified the strain JC1078T in a novel genus and species, *Serinicoccus marinus* gen. nov., sp. nov. Sahu *et al.* (2004) reported 40 strains of actinobacteria from the gut contents of three estuarine fishes *viz. Chanos chanos, Etroplus suratensis* and *Lates calcarifer*. The isolated strains were tested for their antagonistic activity against six bacterial species *viz. B. subtilis, Klebsiella pneumoniae, Proteus vulgaris, S. aureus, Shigella flexneri* and *Vibrio cholerae.* Among them, only 10 strains of actinobacteria (30 per cent) showed moderate antagonistic activity against all the tested pathogens. Magarvey *et al.* (2004) isolated 102 strains of actinobacteria from subtidal marine sediments of Bismarck Sea and the Solomon Sea off the coast of Papua New Guinea. Based on the molecular studies, they identified two strains belonging to two new genera (represented by strains of the PNG1 clade and strain UMM518) within the family Micromonosporaceae. The strains showed activities against multidrug-resistant gram-positive pathogens, malignant cells, and vaccinia virus replication.The structure of a new dibenzodiazepine alkaloid, diazepinomicin (1), isolated from a marine *Micromonospora* sp. was characterized by Charan *et al.* (2004) using spectroscopic methods. Diazepinomicin represents a unique molecular class composed of a dibenzodiazepine core linked to a farnesyl side chain. The ethyl acetate extract from marine *Streptomyces* sp. isolate B8652 delivered the trioxacarcins A to approximately C and additionally three new derivatives designated as trioxacarcins D to approximately F was repoted by Maskey *et al.* (2004). All trioxacarcins showed high anti-bacterial and some of them, high anti-tumor and anti-malaria activity. Structures of the new antibiotics were derived from mass, 1D and 2D NMR spectra and confirmed by comparison of the NMR data with those of known derivatives. The absolute configuration of the trioxacarcins was deduced from the X-ray analysis of gutingimycin (2g) and from the known stereochemistry of L-trioxacarcinoses A and B. Two new caprolactones, (R)-10-methyl-6-undecanolide (1) and (6R, 10S)-10-methyl-6-dodecanolide (2), were identified in the lipid extract of a marine streptomycete (Stritzke *et al.*, 2004). Their structures were proposed on the basis of GC-MS experiments and proved by synthesis. The absolute configuration of the compounds was established by comparison of the natural and synthetic stereoisomers using chiral gas chromatography. These caprolactones showed a moderate phytotoxicity and a promising activity against cancer cells with concomitant low general cytotoxicity.

Lang *et al.* (2004) isolated a *Microbacterium* species from the sponge *Halichondria panicea*, capable of producing a glucosylmannosyl-glycerolipid when grown on a complex medium with glucose. Optimizing the carbon sources in shake flask experiments had shown that glycerol afforded the highest specific glycoglycerolipid production. The product yield approached 300 mg/L or 25 mg/g biomass upon scaling up in a 40-L bioreactor volume. The native diglycosyl-glycerolipid GGL.2 strongly inhibited the growth of the tumor cell lines HM02 and Hep G2 (50 per cent inhibition at 0.4 to 3 microg/mL), while the related deacylated compound (GG.2) showed a potent anti-tumor-promoting activity. Srivibool *et al.* (2004) screened 567 strains of thermotolerant actinobacteria from sixty four soil samples collected from the coastal ecosystem of Chonburi, Rayong and Trat Provinces. The isolated strains were used for assay of *N*-acylamino acid racemase activity and two isolates viz. *Streptomyces tendae* and *S. maritimus* could produce L-methionine as end product at 0.68 mM and 0.21 mM respectively. Activity of *N*-acylamino acid racemase was reported for the first time from these species of *Streptomyces. S. aureoverticillatus* isolated by Mitchell *et al.* (2004) from the marine sediments was found to produce a novel macrocyclic lactam with cytotoxicity against various tumor cell lines. Using extensive MS, UV, and NMR spectral analyses, the structure was elucidated as compound 1, aureoverticillactam, a 22-atom macrocyclic lactam incorporating both triene and tetraene conjugated olefins. A halophilic *Actinopolyspora* species AHI was isolated from the sediments of Alibag coast of Maharastra, India (Kokare *et al.*, 2004a). This strain exhibited good antagonistic activity against gram-positive bacteria and *Trichoderma* sp. It did not show any antibacterial activity against gram-negative bacteria. The strain showed resistance to clindamycin, vancomycin, nalixidic acid and streptomycin antibiotics. Furthermore, Kokare *et al.* (2004b) screened 35 bioactive producing actinobacteria from Maharastra and Goa Coasts, India. Out of the 35 strains, 20 showed strong antimicrobial activity against different types of microorganisms and remaining 15 showed very weak antimicrobial activity.

Ellaiah *et al.* (2004) reported a new bioactive producing *Streptomyces kavutarensis* sp. nov. from the marine sediments of Machilipatnam, India. Kathiresan *et al.* (2005) isolated 160 strains from the sediments of mangrove, estuary, sand dune and industrially polluted marine environment of Cuddalore. Of these, mangrove sediments were the rich sources for actinobacteria. When these isolates were tested against phytopathogenic fungi *viz., Rhizoctonia solani, Pyricularia oryzae, Helminthosporium oryzae* and *Colletotrichum falcatum*, about 51 per cent of the isolates were effective against *P. oryzae* and *H. oryzae*, 31 per cent against *R. solani* and 12.5 per cent against *C. falcatum*. Of the 160 isolates, 10 showed potent activity against all the fungi tested. These isolates produced high antifungal compounds at 120 h of incubation period in the production medium culture. Glucose and soyabean meal were the best carbon and nitrogen sources respectively and 17.5 ppt was the best salinity level for maximum antibiotic production. Sivakumar *et al.* (2005a, b) reported 91 strains of actinobacteria from different stations of the Pichavaram mangrove ecosystem. The isolated strains were tested for their antagonistic activity against the potential human pathogens such as *B. subtilis, P. vulgaris, S. flexneri, K. pneumoniae, V. cholerae* and *S. aureus*. Out of the 91 strains, only 6 strains showed good activity and they were identified upto species level. A set of 600 actinobacterial strains were isolated by Fiedler *et al.* (2005) from the marine sediments of various sites in the Pacific and Atlantic Oceans. They reported a new marine member of the rare genus *Verrucosispora* which produced novel bioactive secondary metabolites (abyssomicin).

Liu *et al.* (2005) isolated *Saccharopolyspora* sp. nov., from the marine sediments which produced two known prodigiosin analogs—metacycloprodigiosin (1) and undecylprodigiosin (2). These compounds exhibited significant cytotoxic activities against five cancer cell lines: P388, HL60, A-549, BEL-7402, and SPCA4. This is the first report on the significant cytotoxicity of metacycloprodigiosin (1) against human cancer cell lines. *S. nodosus* isolated from the marine sediments of Scripps Canyon, La Jolla, California, was found to produce lajollamycin (1), a nitro-tetraene spiro-beta-lactone-gamma-lactam antibiotic (Manam *et al.*, 2005). The structure was established by complete analysis of spectroscopic data and comparison with known antibiotics oxazolomycin (2), 16-methyloxazolomycin (3), and triedimycin B (4). Lajollamycin (1) showed antimicrobial activity against both drug-sensitive and resistant gram-positive bacteria and inhibited the growth of B16-F10 tumor cells *in vitro*. Two novel antitumor antibiotics designated as chinikomycins A (2a) and B (2b) were isolated together with manumycin A (1) from marine streptomycetes (Li *et al.*, 2005). Chinikomycins A (2a) and B (2b) are chlorine-containing aromatized manumycin derivatives of the type 64-pABA-2 with an unusual para orientation of the side chains. The compounds exhibited antitumor activity against different human cancer cell lines, but were inactive in antiviral, antimicrobial, and phytotoxicity tests. Lee *et al.* (2005) reported two new cyclic peptides (2 and 3) along with the nocardamine (1) from *Streptomyces* sp. isolated from an unidentified marine sponge. On the basis of the results of combined spectral analyses, the structures of the new compounds were defined to be the dehydroxy and desmethylenyl derivatives of nocardamine, respectively. The new compounds exhibited weak inhibition against the enzyme sortase B. Buchanan *et al.* (2005) discovered two unprecedented macrolides, sporolides A (1) and B (2) from the marine *Salinispora tropica*. The structures and absolute stereochemistries of both metabolites were elucidated using a combination of NMR spectroscopy and X-ray crystallography. A *Streptomyces* sp. isolated from the marine sediments of Alaska was found to produce three new pyrrolosesquiterpenes, glyciapyrroles A (1), B (2), and C (3), along with the known diketopiperazines cyclo(leucyl-prolyl) (4), cyclo(isoleucyl-prolyl) (5), and cyclo(phenylalanyl-prolyl) (6) (Macherla *et al.*, 2005a). Further, Macherla *et al.* (2005b) reported Salinosporamide A, a potent proteasome inhibitor useful in treating cancer. In this study, a series of analogues was assayed for cytotoxicity, proteasome inhibition, and inhibition of NF-kappaB activation. Marked reductions in potency in cell-based assays accompanied replacement of the chloroethyl group with unhalogenated substituents. Halogen exchange and cyclohexene ring epoxidation were well tolerated, while some stereochemical modifications significantly attenuated activity. These findings have provided insights into structure-activity relationships within this novel series. A new cytotoxic substance named mechercharmycin A was isolated from marine-derived *Thermoactinomyces* sp. (Kanoh *et al.*, 2005). The structure of mechercharmycin A was determined by an X-ray crystallographic analysis to be cyclic peptide-like and bearing four oxazoles and a thiazole. Mechercharmycin B, a linear congener of mechercharmycin A, was also isolated from the same bacterium. Mechercharmycin A exhibited relatively strong antitumor activity, whereas mechercharmycin B exhibited almost no such activity. Yan *et al.* (2005) isolated a total of 354 actinobacteria strains from near-shore marine samples of Wenchang mangrove, DanZhou harbour and YanPu harbour using different isolation methods. Among the 30 strains of streptomycetes showed strong cytotoxic activity against B16 cell.

Soria-Mercado *et al.* (2005) screened actinobacteria from ocean sediments of La Jolla, California and based on the 16S rDNA sequence-based phylogenetic analysis, the strains were recognized as members of a new genus (tentatively called MAR4) within the family Streptomycetaceae. One isolated strain CNQ-525 produced three new chlorinated dihydroquinones (1-3) and one previously reported analogue, 4. The compounds had new carbon skeletons, but were related to several previously reported metabolites of the napyradiomycin class. Structures of the new molecules which possessed significant antibiotic properties and cancer cell cytotoxicities, were assigned by comprehensive spectral measurements and by comparison with NMR and other spectral data from the antibiotic A80915C (5) and the full stereostructure was later assigned by X-ray diffraction methods. Williams *et al.* (2005) isolated an obligate marine *Salinispora tropica* (strain CNB-392), which produced potent proteasome inhibitor salinosporamide A (1) and led to the isolation of seven related gamma-lactams. The most important of these compounds were salinosporamide B (3), which is the deschloro-analogue of 1, and salinosporamide C (4), which is a decarboxylated pyrrole analogue. New SAR data for all the eight compounds, derived from extensive testing against the human colon carcinoma HCT-116 and the 60-cell-line panel at the NCI, indicated that the chloroethyl moiety plays a major role in the enhanced activity of 1. Kock *et al.* (2005) reported a marine-derived streptomycete which produced three known antibiotics, resistomycin (1), resistoflavin (3a) and tetracenomycin (4) and a further member of the rare resistomycin class, the weakly antibiotically active 1-hydroxy-1-norresistomycin (2). Gorajana *et al.* (2005) screened 1-Hydroxy-1-norresistomycin, a new cytotoxic compound from a marine actinobacteria, *Streptomyces chibaensis,* collected from marine sediment samples of the Bay of Bengal, India. The compound showed a potent cytotoxic activity against cell lines *viz.*, HMO_2 (gastric adenocarcinoma) and HePG2 (hepatic carcinoma) *in vitro*. It also exhibited antibacterial activities against gram-positive and gram-negative bacteria. Sujatha *et al.* (2005a) studied antagonistic marine actinobacteria collected from the Bay of Bengal, India. The three isolated strains (No. BT-606, BT-624, BT-652) showed very promising antagonistic activities against multi-drug resistant pathogens. Further, Sujatha *et al.* (2005b) reported a new *Streptomyces psammoticus* from the sediments of Bay of Bengal, India, which produced polyketide antibiotic SBR-22 and showed antibacterial activity against methicillin resistant *Staphylococcus aureus.*

You *et al.* (2005) screened a total of 94 actinobacteria strains from the marine sediments of a shrimp farm and 87.2 per cent strains belonged to the genus *Streptomyces* and others were *Micromonospora* spp. About 50 per cent of the strains showed activity against the pathogenic *Vibrio* spp. Thirty-eight percent of marine *Streptomyces* strains produced siderophores on Chrome Azurol S (CAS) agar plates. Seven strains of *Streptomyces* were found to produce siderophores and inhibited the growth of *Vibrio* spp. *in vitro*. Two of them belonged to the *Cinerogriseus* group, the most frequently isolated group of *Streptomyces*. Thus, the studies showed that streptomycetes could be a promising source for the isolation of biocontrol agents in aquaculture. Several enzyme-inhibitor and antibiotic producing actinobacteria were reported by Imada (2005) from various samples of the marine environment. Most of them produced novel compounds, useful in medicine and agriculture. Actinobacterial strain no. 18, which produce D antibiotics only against gram-positive bacteria in the presence of seawater was isolated from sediments sampled from neritic sea water and characterized. The production of antibiotics was observed at seawater concentrations ranging from 60 to 110 per cent (v/v). The production of tetrodotoxin (TTX), known otherwise as puffer fish toxin, was investigated in various actinobacteria collected from the marine environment. Of 10 the isolates collected from various sea areas, 9 produced TTX as judged by their retention times, using the High Performance Liquid Chromatography (HPLC). Probably, this was the first report of actinobacteria from the marine environment that produced TTX. NPI-3114, chlorinated bisindole pyrrole and NPI-3304, a 4-oxazolidinone (Gravestock, 2005). The three dimensional structure of NPI-3114 was distinct from the classic indolocarbazole, staurosporine, and therefore has distinct biological properties. NPI–3304 was structurally related to the antibiotic, linezolid, an antibacterial compound previously discovered and developed directly from a synthetic chemistry program. NPI-3114 and NPI-3304 showed broad gram-positive antibacterial activity, including activity against methicillin resistant *Staphylococcus aureus,* and no hemolytic or cytotoxicity against normal fibroblasts. Thiocoraline is a thiodepsipeptide antitumor compound produced by two actinobacteria *Micromonospora* sp. ACM2-092 and *Micromonospora* sp. ML1, isolated from two marine invertebrates (a soft coral and a mollusc) found off the Indian Ocean coast of Mozambique (Lombo *et al.*, 2006). By using oligoprimers derived from Nonribosomal Peptide Synthetase (NRPS) consensus sequences, six PCR fragments containing putative NRPS adenylation domains were amplified from the chromosome of *Micromonospora* sp. ML1. Insertional inactivation of each adenylation domain showed that two of them generated nonproducing mutants, thereby indicating that these domains were involved in thiocoraline biosynthesis. Sequencing of a 64.6 kbp DNA region revealed the presence of 36 complete Open Reading Frames (ORFs) and two incomplete ones.

Heterologous expression of a region of about 53 kbp, containing 26 of the ORFs, in *Streptomyces albus* and *S. lividans* led to the production of thiocoraline in these streptomycetes. Surprisingly, the identified gene cluster contains more NRPS modules than expected on the basis of the number of amino acids of thiocoraline. TioR and TioS would most probably constitute the NRPS involved in the biosynthesis of the thiocoraline backbone, according to the colinearity of the respective modules. It was proposed that two other NRPSs, TioY and TioZ, could be responsible for the biosynthesis of a small peptide molecule which could be involved in regulation of the biosynthesis of thicoraline in *Micromonospora* sp. ML1. In addition, a pathway was proposed for the biosynthesis of the unusual starter unit, 3-hydroxy-quinaldic acid. Four antitumor-antibiotics of a new structure class, the marinomycins A-D (1-4), were isolated by Kwon *et al.* (2005) from the saline culture of a new group of marine actinobacteria, *Marinispora* sp. Sahu *et al.* (2006) reported a total number of 40 strains of actinobacteria from the sediments of the Vellar estuary and checked their antagonistic activity against human bacterial pathogens (*B. substilis, Pseudomonus vulgaris, Shigella flexineri, K. pneumoniae, V. cholerae* and *S. aureus*). Among them, 9 strains (22.5 per cent) showed activity against the tested pathogens and 5 strains which showed good activity were identified upto species level. Muthurayar *et al.* (2006) isolated a total of 18 actinobacterial strains from an estuarine fish, *Chanos chanos* and studied their antagonistic activity against human bacterial pathogens. Out of the 18 strains, only five strains *viz.* A-1, AA-5, AA-10, AA-13 and AA-17 (27.7 per cent) showed moderate activity against all the tested pathogens. These strains were mutated using physical (UV radiation) and chemical (NTG) mutagens. After mutation, antibacterial activity of the strains *viz.* AA-13 (UV treated) and AA-10 (NTG treated) was found. But the strain AA-17 (UV treated and NTG treated) showed decreased antibacterial activity against all the tested human pathogens. The study suggested that mutation is one of the good methods for strain development to increase the efficiency of the actinobacteria for antibacterial activity.

Oh *et al.* (2006) reported two structurally novel cyclopenta[a]indene glycosides, cyanosporasides A and B (1 and 2) from the culture broth of a new species of the obligate marine actinomycete genus *Salinispora*. The cyanosporasides possessed a new 3-keto-pyranohexose sugar as well as a cyano- and chloro-substituted cyclopenta[a]indene ring system. The cyanosporasides were proposed to be cyclization products of an enediyne precursor. Two new polyketides, actinofuranones A (1) and B (2), were screened by Cho *et al.* (2006) from the culture extract of a marine-derived *Streptomyces* strain (CNQ766). The relative stereochemistries of these new molecules were assigned on the basis of analysis of NOE data and vicinal (1)H-(1)H coupling constants, while the absolute configurations of the asymmetric centers were determined using the modified Mosher's method. Gorajana *et al.* (2006) isolated a bioactive producing *Streptomyces chibaensis* from the marine sediment samples of Bay of Bengal, India. The isolate yielded a cytotoxic compound *i.e.* quinone-related antibiotic, resistoflavine (1) which showed potent cytotoxic activity against cell lines *viz.*, HMO_2 (Gastric adenocarcinoma) and HePG2 (Hepatic carcinoma) *in vitro* and also exhibited weak antibacterial activities against gram-positive and gram-negative bacteria. Peng *et al.* (2006) reported two actinobacteria, *Streptomyces* NRRL 5690 and *Streptomyces spheroides*, which significantly metabolized cyanthiwigin B to new metabolites. *Streptomyces* NRRL 5690 transformed cyanthiwigin B to three new compounds, cyanthiwigins AE (2), AF (3), and AG (4) and the known cyanthiwigin R (5). *S. spheroides* transformed cyanthiwigin B to cyanthiwigins S (6), E (7), and AE (2). All microbial-metabolized derivatives (2-7) of cyanthiwigin B exhibited the ability to increase the antimicrobial activity of curcuphenol, the major antimicrobial sesquiterpene isolated from *M. styx*. Adinarayana *et al.* (2006) isolated *Strepomyces corchorusii* a bioactive species from the marine sediment samples collected from the Bay of Bengal, India, which produced two cytotoxic compounds *viz.*, resistomycin, a quinone-related antibiotic, and tetracenomycin D, an anthraquinone antibiotic. They showed potent cytotoxic activity against cell lines HMO2 (gastric adenocarcinoma) and HepG2 (hepatic carcinoma) and also exhibited weak antibacterial activities against gram-positive and gram-negative bacteria. A new cytotoxic compound, streptokordin, and four known compounds, nonactic acid, dilactone, trilactone, and nonactin, were isolated from a marine *Streptomyces* collected from deep-sea sediments (Jeong *et al.*, 2006). This strain produced cytotoxic compound, designated as streptokordin. Its structure was elucidated by FAB-MS, 1H, 13C, and 2D NMR spectroscopy. Streptokordin exhibited significant cytotoxicity against seven human cancer cell lines but showed no growth inhibition against various microorganisms including bacteria and fungi. Wu *et al.* (2006) screened a new staurosporinone, N-carboxamido-staurosporine (1c), and a new sesquiterpene, (5S,8S,9R, 10S)-selina-4(14),7(11)-diene-8,9-diol (2a), from the marine-derived *Streptomyces* sp. QD518. Four known butenolides from the marine-derived bacterium, *Streptoverticillium luteoverticillatum* were reported by Li *et al.* (2006). The compounds showed *in vitro* cytotoxicity against the murine lymphoma P388 and human leukemia K562 cell lines. Hongxia *et al.* (2006) obtained a marine actinobacteria isolate M097 from the marine sediments in Jiaozhou Bay, China, which produced two anthraquinone compounds *viz.*, loesaponarin II and 1,6-dihydroxy-8-hydroxymethylanthraquinone. It was the first report that the latter had been found as a secondary metabolite from a wild-type strain.

Conclusion

From the above information it shows that the marine actinobacteria are cable to produce different kinds of antibiotic and cytotoxic compounds which have diverse application in the field of medical and agriculture. As the most of the antibiotic and cytotoxic compounds which are marketed are now coming from the terrestrial actinobacteria and only few from the marine actinobacteria. So there is need a lot of research and trial so that compounds from the marine actinobacteria could be patented and marketed as our Ocean is covered 75 per cent of earth surface which harbour diverse and potential actinobacteria.

Acknowledgement

Authors thank Prof. T. Balasubramanian, Director, Centre of Advanced Study in Marine Biology and the authorities of Annamalai University for providing with necessary facilities. One of the two authors are thankful to the Ministry of Environments and Forests, Government of India for granting the financial support in form of fellowship.

References

Adinarayana, G., M.R. Venkateshan, V.V. Bapiraju, P. Sujatha, J. Premkumar, P. Ellaiah and A. Zeeck, 2006. Cytotoxic compounds from the marine actinobacterium. *Bioorg. Khim.*, 32: 328-334.

Asolkar, R.N., R.P. Maskey, E. Helmke and H. Laatsch, 2002. Chalcomycin B, a new macrolide antibiotic from the marine isolated *Streptomyces* sp. B7064. J. *Antibiot.*, 55: 893-898.

Balagurunathan, R. and A. Subramanian, 2001. Antagonistic streptomycetes from marine sediments. *Ad. Bios.*, 20: 71-76.

Balagurunathan, R., G.S. Prasad, R. Manavalan and A. Subramanian, 1989. Actinomycetes from the littoral sediments of Parangipettai (South India) and their antibiotic activity. In: Proceeding of the

First International Marine Biotechnology Conference, Centre of Advanced Study in Marine Biology, Annalmalai University, Parangipettai, India, p 10.

Bernan, V.S., D.A. Montenegro, W.M. Maiese, D.A. Steinberg and M. Greenstein, 1994. Bioxalomycins, new antibiotics produced by the marine *Streptomyces* sp.LL-31F508: taxonomy and fermentation. *J. Antibiot.*, 47: 1417-1424.

Biabani, M.A., H. Laatsch, E. Helmke and H. Weyland, 1997. Delta-Indomycinone: a new member of pluramycin class of antibiotics isolated from marine *Streptomyces* sp. *J. Antibiot.*, 50: 874-877.

Buchanan, G.O., P.G. Williams, R.H. Feling, C.A. Kauffman, P.R. Jensen and W. Fenical, 2005. Sporolides A and B: structurally unprecedented halogenated macrolides from the marine actinomycete *Salinispora tropica*. *Org. Lett.*, 7: 2731-2734.

Canedo, L.M., J.L. Fernandez-Puentes and J.P. Baz, 2000. IB-96212, a novel cytotoxic macrolide produced by a marine *Micromonospora*. II. Physico-chemical properties and structure determination. *J. Antibiot.*, 53: 479-483.

Capon, R.J., C. Skene, E. Lacey, J.H. Gill, J. Wicker, K. Heiland and T. Friedel, 2000. Lorneamides A and B: two new aromatic amides from a Southern Australian marine actinomycete. *J. Nat. Prod.*, 63: 1682-1683.

Charan, R.D., G. Schlingmann, J. Janso, V. Bernan, X. Feng and G.T. Carter, 2004. Diazepinomicin, a new antimicrobial alkaloid from a marine *Micromonospora* sp. *J. Nat. Prod.*, 67: 1431-1433.

Cheng, X. C., P.R. Jensen and W. Fenical, 1999. Luisols A and B, new aromatic tetraols produced by an estuarine marine bacterium of the genus *Streptomyces* (Actinomycetales). *J. Nat. Prod.*, 62: 608-610.

Cho, J.Y., H.C. Kwon, P.G. Williams, C.A. Kauffman, P.R. Jensen and W. Fenical, 2006. Actinofuranones A and B, polyketides from a marine-derived bacterium related to the genus *Streptomyces* (acttinomycetales). *J. Nat. Prod.*, 69: 425-428.

Cho, K.W., H.S. Lee, J.R. Rho, T.S. Kim, S.J. Mo and J. Shim, 2001. New lactone-containing metabolites from a marine-derived bacterium of the genus *Strptomyces*. *J. Nat.Prod.*, 64: 664-667.

Dhevendaran, K. and Y.K. Anithakumari, 2002. L-asparaginase activity in growing conditions of *Streptomyces* spp. associated with *Therapon jarbua* and *Villorita cyprinoids* of Veli Lake, South India. *Indian J. Mar. Sci.*, 39: 155-159.

Dixit, V.S. and A. Pant, 2000. Comparative characterization of two serine endopeptidases from *Nocardiopsis* sp. NCIM 5124. *Biochim Biophys Acta.*, 18 (2-3): 261-268.

Ellaiah, P., D. Kalyan, V.S. Rao and B.V. Rao, 1996. Isolation and characterization of bioactive actinomycetes from marine sediments. *Hiodustan Antibiot. Bull.*, 38 (1-4): 48-52.

Ellaiah, P., G. Adinarayana, J.P. Kumar, V. Saisha and P. Vasu, 2004. Characterization of a new bioactive actinomycete AUB N5/8 from marine sediments. *Hidustan Antiiotic Bull.*, 46 (1-4): 16-21.

Feling, R.H., G.O.Buchanan, T.J. Mincer, C.A. Kauffman, P.R. Jensen and W. Fenical, 2003. Salinosporamide A: a highly cytotoxic proteasome inhibitor from a novel microbial source, a marine bacterium of the genus *Salinospora*. *Angew. Chem. Int. Ed. Engl.*, 42 (3): 355-357.

Fernandez-Chimeno, R.L., L. Canedo, F. Espliego, D. Gravalos, F. De La Calle, J.L. Fernandez-Puentes and F. Romero, 2000. IB-96212, a novel cytotoxic macrolide produced by a marine *Micromonospora*. I. Taxonomy, formation, isolation and biological activities. *J. Antibiot.*, 53 (5): 474-478.

Fiedler, H.P., C. Bruntner, A.T. Bull, A.C. Ward, M. Goodfellow, O. Potterat, C. Puder and G. Mihim, 2005. Marine actinomycetes as a source of novel secondary metabplites. *Antonie Van Leeuwenhoek*, 87 (1): 37-42.

Goodfellow, M. and A.G. O'Donnell, 1989. Roots of Bacterial Systematic. *In* The Handbook of new bacterial systematic. Academic Press, London, pp. 3-54.

Goodfellow, M. and J.A. Haynes, 1984. Actinomycetes in marine sediments. *In* Biological, biochemical and biomedical aspects of actinem. Edited by I. Ortiz-Ortiz, L.F. Bojalil and V. Yakoleff). Academic Press, London, pp. 453-472.

Goodfellow, M. and S.T. Willams, 1983. Ecology of actinomycetes. *Ann. Rev. Microbiol.*, 37: 189-216.

Gorajana, A., B.V. Kurada, S. Peela, P, Jangam, S. Vinjamuri, E. Poluri and A. Zeeck, 2005. 1-Hydroxy-1-norresistomycin, a new cytotoxic compounds from a marine actinomycete, *Streptomyces chibaensis* AUBN1/7. *J. Antibiot.*, 58 (8): 526-529.

Gorajana, A., S. Vinjamuri, B. V. Kurada, S. Peela, P. Jangam and E. Poluri, 2006. Resistoflavine, cytotoxic compound from a marine actinomycete, Streptomyces chibaensis AUBN(1)/7. *Microbiol. Res.*, 26: 512-518.

Gravestock, M.B., 2005. Recent developments in the discovery of novel oxazolidinone antibacterials. *Curr. Opin. Drug Discov. Devel.*, 8(4):469–477.

Hernan, L.M., J.A. Blanco, J.P. Baz, J.L. Puentes, F.R. Millan, F.E. Vazquez, R.I. Fernandez-Chimeno and D.G. Gravalos, 2000. 4'-N-methyl-5'-hydroxystaurosporine and 5'-hydroxystaurosporine, ne indolocarbazole alkaloids from a marine *Micromonospora* sp. strain. *J. Antibiot.*, 53 (9): 895-902.

Hernandez, I.L., M.J. Godinho, A. Magalhaes, A.B. Schefer, A.G. Ferreira and R.G. Berlinck, 2000. N-acetyl-gamma-hydroxyvaline lactone, an unusual amino acid derivative from a marine streptomycete. *J. Nat. Prod.*, 63(5): 664-665.

Hertwecka, C. and B.S. Moore, 2000. A Plant-like Biosynthesis of Benzoyl-CoA in the Marine Bacterium *Streptomyces maritimus*. *Tetrahedron*, 56: 9115-9120.

Hotta, K., M. Yoshida, M. Hamada and Y. Okami, 1980. Studies on new aminoglycoside antibiotics, istamycins, from an actinomycetes isolated from a marine environment. III. Nutritional effects on istamycin production and additional chemical and biological properties of istamycins. J. *Antibiot.*, 33 (12): 1515-1520.

Itoh, T., M. Kinoshita, H. Wei and M. Kobaysahi, 2003a. Stereostructure of komodoquinone A, a neuritogenic anthracycline, from marine *Streptomyces* sp. KS3. *Chem. Pharm. Bull.*, 51 (12): 1402-1404.

Itoh, T., M. Kinoshita, S. Aoki and M. Kobaysahi, 2003b. Komoduinoone A, a novel neuritogenic anthracycline, from *Streptomyces* sp. KS3. *J. Nat. Prod.*, 66: 1373-1377.

Ivanova, V., M. Oriol, M.J. Montes, A. Garcia and J. Guinea, 2001. Secondary metabolites from a *Streptomyces* strain isolated from Livingston Island. Z. *Naturoforsch*, 56 (1-2): 1-5.

Jensen, P.R., R. Dwight and W. Fenical, 1991. Distribution of actinomycetes in near-shore tropical sediments. *Appl. Environ. Microbiol.*, 57 (4): 1102-1108.

Jeong, S.Y., H.J. Shin, T.S. Kim, H.S.Lee, S.K. Park and H.M. Kim, 2006. Streptokordin, a new cytotoxic compound of the methylpyridine class from a marine-derived Streptoyces sp. KORDI-3238. *J. Antibiot (Tokyo)*, 59 (4): 234-240.

Jrumg, Z.D., P.R. Jensen and W. Fenieal, 1997. Actinellavosiae, A novel flavonoid like glycoside produced by a marine bacterium of the genus *Streptomyces. Tetra. Lett.*, 38 (29): 5605-5608.

Kanoh, K., Y. Matsuo, K. Adachi, H. Imagawa, M. Nishizawa and Y. Shizuri, 2005. Mecherchamycins A and B, cytotoxic substances from marine-derived Thermoactinomyces sp. YM3-251. *J. Antibiot.*, 58(4): 289-292.

Kathiresan, K., R. Balagurunathan and M. Masilamani Selvam, 2005. Fungicidal activity of marine actinomycetes against phytopathogenic fungi. *Indian J. Biotech.*, 4: 271-276.

Kock, I., R.P. Maskey, M.A.Biabani, E. Helmke and H. Laastsch, 2005. 1-Hydroxy-1-norresistomycin and ressistoflavin methyl ether: new antiotics from marine-derived streptomycetes. *J. Antibiot.*, 58(8): 530-534.

Kokare, C.R., K.R. Mahadik and S.S. Kadam, 2004. Isolation of bioactive actinomycetes from sediments from Goa and Maharastra coastlines (West coast of India). Indian *J. Mar. Sci.*, 33 (3): 248-256.

Kokare, C.R., K.R. Mahadik, S.S. Kadam and B.A. Chopada, 2004a. Isolation, characterization and antibacterial activity of marine halophilic *Actinopolyspora* species AH1 from the west coast of India. *Curr. Sci.*, 86 (4): 593-597.

Kuster, E., 1976. Ecology and predominance of soil strerptomycete. In Actinomycetes–The boundary microorganisms. Edited by T. Arai. Tokyo, Toppan Co. Ltd., pp. 109-113.

Kwon, H.C., C.A. Kauffman, P.R. Jensen and W. Fenical, 2006. Marinomycins A-D, antitumor-antibiotics of a new structure class from a marine actinomycetes of the recentlty discovered genus *Marinispora. J. Am. Chem. Soc.*, 128 (5): 1622-1632.

Laksmanaperumalsamy, P., D. Chandramohan and R. Natarajan, 1978. Antibacterial and antifungal activity of streptomycetes from Porto Novo coastal environment. *Mar. Biol.*, 11: 15-24.

Lang, S., W. Beli, H. Tokuda, C. Wicke and V. Lurtz, 2004. Improved production of bioactive glycosylmannosyl-glycerolipid by sponge-associated *Microbacterium* species. *Mar. Biotechnol.*, 6(2): 152-156.

Lechevalier, H., 1982. The development of applied microbiology at Rutgers, The State University of New Jersay.

Lee, H.S., H.J. Shin, K.H. Jang, T.S. Kim, K.B. Oh and J. Shin, 2005. Cyclic peptides of the nocardamine class from a marine-derived bacterium of the genus *Streptomyces. J. Nat. Prod.*, 68(4): 623-625.

Li, D.H., T.J. Ahu, H.B.Liu, Y.C. Fang, Q.Q.Gu and W.M. Zhu, 2006. Four butenolides novel cytoxic compounds isolated from the marine-derived bacterium, *Streptoverticillium luteoverticillatum. Arch. Pharm. Res.*, 29(8) 624-66.

Li, F., R.P. Maskev, S. Qin, I, Sattler, H.H. Fiebig, A. Maier, A. Zeeck and H. Laatsch, 2005. Chinikomycins A and B: islation, structure elucidation and biological activity of novel antibiotics from a marine *Streptomyces* sp. isolated M045. *J, Nat. Prod.*, 68(3): 349-353.

Liu, R. C.B. Cui, L. Duan, Q.Q. Gu and W.M. Zhu, 2005. Potent in vitro anticancer activity of metacycloprodigiosin and undecy;prodigiosin from a sponge-derived actinomycete Saccharopolyspora sp. nov. *Arc. Pharm. Res.*, 28 (12): 1641-1344.

Lombo, F., A. Velasco, de la Calle F., A.F. Brana, J.M. Sanchez-Pulles, C. Mendez and J.A. Salas, 2006. Deciphering the biosynthesis of the antitumor thiocoraline from a marine actinomycete and its expression in two *Streptomyces* species. *Chembiochem*, 7 (2): 366-376.

M.K. Sahu, K. Sivakumar and L. Kannan, 2007. Alkaline protease production by an actinomycete isolated from tiger shrimp, *Penaeus monodon* (Fabricius, 1798). *Natl. Acad. Sci. Lett.*, 30 (1&2): 61-65.

Macherla, V.R., J. Liu, C. Bellows, S. Teisan, B. Nicholson, K.S. Lam and B.S. Potts, 2005a. Glaciapyrroles A, B and C, pyrrolosesquiterpenes from a *Streptomyces* sp. isolated from an Alaskan marine sediment. *J. Nat. Prod.*, 68 (5): 780-783.

Macherla, V.R., S.S. Mitchell, R.R. Manam, K.A., Reed, T.H. Chao, B. Nicholson, G. Devanat-Yazdi, B. Mai, P.R. Jensen. W.F. Fenical, S.T. Neuteboom and K.S. Lam, 2005b. Structure-activity relationship studies of salinosporamide A (NPI-0052), a novel marine derived proteasome inhibitor. *J.Med. Chem.*, 48 (11): 3684-3687.

Magarvey, N.A., J.M. Keller, V. Bernan, M. Dworkin and D.H. Sherman, 2004. Islation and characterization of novel marine-derived actinomycete taxa rich in bioactive metabolites. *Appl. Environ. Microbiol.*, 70 (12): 7520-7529.

Manam, R.R., S. Teisan, D.J. White, B. Nicholson, J. Grodberg, S.T. Neuteboom, K.S. Lam, D.A. Mosca, G.K. Lloyd and B.C. Potts, 2005. Lajollamycin, a nitro-tetraene spiro-beta-lactone-gamma-lactam antibiotic from the marine actinomycete *Streptomyces nodosus*. *J. Nat. Prod.*, 68 (2): 240-243.

Maskey, R.P., E. halmke, O. Kayser, H.H. Fiebig, A. Maier, A. Busche and H. Laatsch, 2004. Anti-cancer and antibacterial trioxacarcins with anti-malaria activity from a marine streptomycete and their absolute stereochemistry. *J. Nat. Prod.*, 57 (12): 771-779.

Maskey, R.P., E. Helmke, H. Laastsch, 2003. Himalomycin A and B: isolation and structure elucidation of new fridamycin type antibiotics from a marine *Streptomyces* isolate. *J. Antibiot.*, 56 (11): 942-949.

Maskey, R.P., E. Helmke, H.H. Fiebig and H. Laatsch, 2002. Parimycin: isolation and structure elucidation of a novel cytotoxic 2,3-dihdroquinizarin analogue of gamma-indomycinone from a marine *Streptomyces* isolate. *J. Antbiot.*, 55 (12): 1031-1035.

Maskey, R.P., F. Li, S. Qin, H.H. Fiebig and H. Laatsch, 2003. Chandrananimycins A approximately C: production of novel anticancer from a marine *Actinomadura* sp. isolate M048 by variation of medium composition and growth condition. *J. Antibiot.*, 56 (7):622-629.

Mathew, A. and R. Philip, 2003. Marine actinomycetes as antagonistic agents to bacterial prawn pathogens. In Aquaculture medicine. Edited by B.I.S. Singh, S. Somnath Pai, R. Philip and A. Monandas. Center for Fish Discasc Diagnostic and Management, School of Environmental Studies, Cochin University of Science and Technology, India, pp. 69-79.

Mathew, A., K. Dhevendaran, M. Georgekutty and P. Natarajan, 1994. L-asparaginase activity in antagonistic streptomycetes associated with clam *Villorita cyprinoides* (Hanley). *Indian J. Mar. Sci.*, 23: 204-208.

Mitchell, S.S., B. Nicholson, S. Teisan, K.S. Lam and B.C. Potts, 2004. Aureoverticillactam, a novel 22-atom macrocyclic lactam from the marine actinomycete *Streptomyces aureoverticillatus*. *J. Nat. Prod.*, 64 (8): 1400-1402.

Moore, B.S., Jacqueline A. Trischman, Dieter Seng, David Kho, Paul R. Jensen and William Fenical, 1999. Salinamides, Antiinflammatory Depsipeptides from a Marine streptomycete. *J. Org. Chem.*, 64: 1145-1150.

Mukku, V.J., M. Speitling, H. Laatsch and E. Helmke, 2000. New butenolides from two marine streptomycetes. *J. Nat. Prod.*, 63 (11): 1570-1572.

Murugan M., M. Srinivasan, K. Sivakumar, Maloy Kumar Sahu and L. Kannan, 2007. Characterization of an actinomycete isolated from the estuarine finfish, *Mugil cephalus* Lin. (1758) and its optimization for cellulase production. *J. Sci. Ind. Res.*, 66 (5): 388-393.

Muthurayar T, K. Sivakumar, M.K. Sahu, T. Thangaradjou and L. Kannan, 2006. Mutational effect on the antibacterial activity of marine actinomycetes isolated from *Chanos chanos*, (Forskal, 1775). *Environ. Eco.*, 24 (1): 46-50.

Oh, D.C., P.G. Williams, C.A. Kauffman, P.R. Jensen and J. Fenical, 2006. Cyanosporasides A and B, chloro and cyano-cyclopenta(a)indene glycosides from the marine actinomycete "Salinispora pacifica". *Org. Lett.*, 8(6): 1021-1024.

Okami, Y., 1979. Antibiotics from marine microorganisms with reference to plasnmid involvement. *J. Nat. prod.*, 42 (6): 583-595.

Okami, Y., 1984. Marine microorganisms as a source of bioactive agents In Current Perspective in Microbial ecology. Edited by M.J. Klug and C.A. Reddy. American Society for Microbiology, Washington DC, pp. 615–655.

Okami, Y., T. Okazaki and T. Kitahara, 1976. Studies on marine microorganisms. V. A new antibiotic, aplasmomycin produced by a streptomycete isolated from shallow shallow sea mud. *J. Antibiot.*, 29 (10): 1019-1025.

Okazaki, T. and Y. Okami, 1972. Studies on marine microorganisms. II. Actinomycetes in Sagami Bay and their antibiotic substances. *J. Antibiot.*, 25 (8): 461-466.

Okazaki, T., T. Kitahara and Y. Okami, 1975. Studies on marine microorganisms. IV. A new antibiotic SS-228 Y produced by *Chainia* isolated from shallow sea mud. J. *Antibiot.*, 28 (3): 176-184.

Patil, R., G. Jeyaskaran and S.A. Shanmugan, 2001b. Occurrence and activity of marine actinomycetes against shrimp bacterial pathogens. *Appl. Fish Aqua.*, 1 (1): 79-81.

Patil, R., G. Jeyaskaran, S.A. Shanmugan and R.J. Shakila, 2001a. Control of bacterial pathogens, associated with fish diseases, by antagonistic marine actinomycetes isolated from marine sediments. *Indian J. Mar. Sci.*, 30 (4): 264-267.

Peng, J., N. Kasanah, C.E Stanley, J. Chadwick, F.R. Fronczek and M.T. Hamann, 2006. Microbial metabolism studies of cyanthiwigin B and synergetic antibiotic effects. *J. Nat. Prod.*, 69 (5): 727-730.

Pisano, M.A., M.J. Sommer and L. Brancaccio, 1989. Isolation of bioactive actinomycetes from marine sediments using rifampicin. *Appl. Microbiol. Biotechnol.*, 31: 609-612.

Pisano, M.A., Michael J. Sommer, and Madelyn M. Lopez, 1986. Application of pretreatments for the isolation of bioactive actinomycetes from marine sediments. *Appl. Microbiol. Biotechnol.*, 25:285-288.

Pusecker, K., H. Laatsch, E. Helmke and H. Weyland, 1997. Dihydrophencomycin methyl ester, a new phenazine derivative from a marine Streptomycete. *J. Antibiot.*, 50 (6): 479-783.

Romero, F., F. Espliego, J. Perez Baz, T. Garcia de Quesada, D. Gravalos, F. de la Calleand J.L. Fernandz-Puentes, 1997. Thiocoraline, a new depsipeptide with antitumoe activity produced by a marine Micromonospora. I. Taxonomy, fermentation, isolation and biological activities. *J.Antibiot.*, 50(9): 734-737.

Sahu, M.K., E. Poorani, K. Sivakumar, T. Thangaradjou and L. Kannan, 2007. Partial purification and anti-leukemic activity of L-asparaginase enzyme of the actinomycete strain LA-29 isolated from an estuarine fish, *Mugil cephalus* (Leinnaeus, 1758). *J. Environ. Biol.*, 8 (3):645-650.

Sahu, M.K., K. Sivakumar and L. Kannan, 2004. Estuarine fish as a source of antagonistic actinomycetes: An inventory. In: Proc Con Microbio Tropi Sea NIO, Goa, MB (O) 04.

Sahu, M.K., K. Sivakumar and L. Kannan, 2006. Isolation and characterization actinomycetes inhibitory to human pathogens. *Geobios*, 33 (2-3): 105-109.

Sahu, M.K., K. Sivakumar, E. Poorani, T. Thangaradjou and L. Kannan, 2007. Studies on L-asparaginase enzyme of actinomycetes isolated from estuarine fishes. *J. Environ. Biol.*, 28 (2/Suppl.): 465-474.

Schumacher, W. R., B.L. Harriganb and B.S. Davidsona, 2001. Kahakamides A and B, new neosidomycin metabolites from a marine-derived actinomycete.*Tetra. Lett.*, 42: 5133–5135.

Shin, J., Y. Seo,H.s. Lee, J.R. Rho and S.J. Mo, 2003. A new cyclic peptide from a marine-derived bacterium of the genus *Nocardiopsis*. *J. Nat. Prod.*, 66 (6): 883-884.

Sitachitta, N., Madhavi Gadepalli, and Bradley S. Davidson, 1996. New a-Pyrone-containing metabolites from a marine-derived actinomycete. *Tchahcdron*, 52(24): 8073-8080.

Sivakumar, K., M.K. Sahu and K. Kathiresan, 2005a. Isolation and characterization of streptomycetes, producing antibiotic, from a mangrove environment. Asian J. Microbiol. *Biotech. Envi. Sc.*, 7 (3): 87-94.

Sivakumar, K., M.K. Sahu and K. Kathiresan, 2005b. An antibiotic producing marine *Streptomyces* from the Pichavaram mangrove environment. *J. Annamalai Uni.*, Part-B, XLI: 9-18.

Sivakumar, K., M.K. Sahu, P.R. Manivel and L. Kannan, 2006. Optimum conditions for L-glutaminase production by actinomycete strain isolated from estuarine fish, *Chanos chanos* (Forskal, 1775). *Indian J. Exp. Biol.*, 44 (March): 256-258.

Skarbek, J.D. and L.R. Brady, 1975. Isolation of Chromomycin A3 from a new subspecies of *Streptomyces*. *Lloydia*, 38 (5): 369-377.

Soria-Mercado, *I.E.*, A. Prieto-Davo, P.R. Jensen and W. Fenical, 2005. Antibiotic terpenoid chloro-dihydroquinones from a new marine actinomycete. *J. Nat. Prod.*, 68(6): 904-910.

Srivibool, R., Kimiko Kurakami, Morakot Sukchotiratanac and Shinji Tokuyamab, 2004. Coastal soil actinomycetes: Thermotolerant strains producing N-Acylamino acid racemase. *Science Asia*, 30:123-126.

Stritzke, K., S. Schulz, H. Laatsch, E. Helmke and W. Beil, 2004. Novel caprolactones from a marine streptomycete. *J. Nat. Prod.*, 67(3): 395-401.

Sujatha, P., K.V.V.S.N. Bapi Raju, T. Ramana, 2005a. Studies on a new marine streptomycete BT-408 producing polyketide antibiotic SBR-22 effective against methicillin resistant *Staphylococcus aureus*. *Microbiol. Res.*, 160: 119-126.

Sujatha, P., V.V.S.N.B. Kurada and R. Terli, 2005b. Studies on antagonistic marine actinomycetes from the Bay of Bengal. *World J. Microb. Biot.*, 21:583–585.

Takaizawa, M, W. Colwell and R.T. Hill, 1993. Isolation and diversity of actinomycetes in the Chesapeake Bay. *Appl. Environ. Microbiol.*, 59: 997-1002.

Tanaka, Y. and O. Omura, 1990. Metabolisms and products of actinomycetes–An Introduction. *Actinomycetologica*, 4:13-14.

Umamaheswary, K., M.K. Sahu, K. Sivakumar, T. Thangaradjou, D. Sumitha and L. Kannan, 2005. Investigations on L-glutaminase producing actinomycetes strain LG-33 from the estuarine fish, *Mugil cephalus* (Linnaeus, 1758). *Environ. Eco.*, 23 (4): 942-947.

Vanajakumar, S., N. Selvakumar and R. Natarajan, 1991. Antagonistic properties of actinomycetes isolated from molluscs of the Porto Novo region, south India. In Bioactive compounds from marine organisms with emphasis on the Indian Ocean. Edited by T. Mary-Frances, R. Sarojini and R. Nagabhushanam. Oxford and IBM Publishing Co. Pvt. Ltd., New Delhi, Bombay and Calcutta, pp. 289-300.

Wicke, C., M. Huners, V. Wray, M. Nimtz, U. Bilitewski and S. Lang, 2000. production and structure elucidation of glycoglycerolipids from a marine sponge- associated Micrbacterium species. *J. Nat. Prod.*, 63(5): 621-626.

Williams, D.E., V.S. Beman, F.V. Ritacco, W. Maiese, M. Greenstein and R.J. Andersen, 1999. Holyrines A and B, possible intermediates in staurosporine biosynthesis produced in culture by a marine actinomycete obtained from the North Atlantic Ocean, *Tetrahedron Lett.*, 40: 7171-7174.

Woo, J.H., E. Kitamura, H. Myouga and Y. Kamei, 2002. An antifungal protein from the marine bacterium *Streptomyces* sp. strain AP77 is specific for *Pythium porphyrae*, a causative agent of red rot disease in *Porphyra* spp. *Appl. Environ. Microbiol.*, 68 (6): 2666-2675.

Wu, S.J., S. Fotso, F. Li, S. Qin, G. Kelter, H.H.Fiebig and H. Laatsch, 2006. N-carboxamido-staursporine and selina-4(14), 7(11)-diene-8,9-diol, new metabolites from a marine *Streptomyces* sp.

Xiong, L., J. Li and F. Kong, 2004. *Streptomyces* sp. 173, an insecticidal microorganisms from marine. *Lett. Appl. Microbiol.*, 38 (1): 32-37.

Xiong, L.X., J.Z. Li and H.L Wang, 2005. *Streptomyces avermitilis* from marine. *J. Environ. Sci.*, 17 (1): 123-125.

Yamada, T., Katsuhiko Minoura and Atsushi Numata, 2002. Halichoblelide, a potent cytotoxic macrolide from a *Streptomyces* species separated from a marine fish. *Tetrahedron Lett.*, 43:1721–1724.

Yan, L.P., K. Hong, S.C. Hu and L.H. Liu, 2005. 16S rDNA diversity analysis of 30 Streptomycetes isolates displaying significant cytotoxic activity against B16 cell from near-shore sediments of Hainan Island. *Wei Sheng Wu Bao*, 45 (2): 185-190.

Yi, H., P. Schumann, K. Sohn and J. Chun, 2004. Serinmicoccus marinus gen. nov., sp. nov., a novel actinomycete with L-aonithine and L-serine in the peptidoglycan. *Int. J. syst. Evol. Microbiol.*, 54: 1585-1589.

You, J.L., L.X. Cao, G.F. Liu, S.N. Zhou, H.M. Tan and Y.C. Lin, 2005. Isolation and characterization of actinomycetes antagonistic to pathogenic, *Vibrio* spp. from nearshore marine sediments, *World J. Microb. Biot.*, 21:679–682.

Chapter 7

Physico-chemical Factors of the Waters of Vattambakkam Lake in Kanchipuram District, Tamil Nadu, India

R. Eswaralakshmi, J. Jayanthi and M.G. Ragunathan

Introduction

The ground water in the earth faces numerous anthropogenic activities which has threatened environmental health of both surface and ground water. Water is a gift to man by nature. But man's activities have made water as a source of potential danger to his health and very existence. The menace of water–borne diseases and epidemics still threatens the well-being of population, particularly in under developed and developing countries. Thus, the quality as well as the quantity of clean water supply is of vital significance for the welfare of man kind (Anon, 1996).

Numerous anthropogenic activities like disposal of sewage and industrial water, recreational activities, excessive usage of fertilizers to lands and use of pesticides has threatened environmental health of both surface and ground water. In addition to this, due to the uneven distribution of fresh water on earth, due to developmental thrust of man and due to difference in the per capital supply and per capita availability of water, the water has become scarce natural resource and national wealth. Water pollution is now days considered not only in terms of public health but also in terms of its conservation, aesthetics and preservation of natural beauty and resources. Water pollution has however threatened to reduce the quantity available in ponds, lakes, rivers and reservoirs due to disposal of sewage, industrial water and due to other human activities (Trivedy and Chandrasekar, 1999).

Population explosion, industrialization, urbanization and developmental thrust of man have created these problems of water pollution. According to different surveys, 70 to 80 per cent of Indian water sources are polluted and different enteric diseases affect million of the peoples every year.

Today most of the water bodies of the world receive million of litres of sewage. Regular drinking water monitoring not only prevents disease and hazards but also checks the water resources from being further polluted (Meitei, *et al.*, 2004).Hence to check the water quality the present investigation has been undertaken.

Materials and Methods

Vattambakkam Lake is rich in variety of crustacean species and fishes. The maximum width of the lake is approximately about 2 miles. The lake is 40 kms away from the Chennai city in Kanchipuram district of Tamil Nadu. It is situated between Padappai and Oragadam. The average depth of the lake is 1.5 m.

Water samples from the Vattambakkam Lake and Guru Nanak College campus were collected in sterilized two liters water bottles and brought to the laboratory for physico-chemical analysis. The characteristics such as appearance, odour, turbidity, total dissolved solids, electrical conductivity, pH, alkalinity, total hardness, calcium, magnesium, sodium, potassium, iron, manganese, free ammonia, nitrite, nitrate, chloride, fluoride, sulphate, phosphate, Tidy's test, silica, biological oxygen demand were estimated in the laboratory by using standard methods (APHA, 1989).

Each sample was tested for ten times and the samples were tested periodically over a period of six months during the year 2007 (January-June) and the results were statistically analyzed and presented.

Results

Water samples collected from Vattambakkam Lake (Sample 1) and the Guru Nanak College campus (Sample 2) were analyzed for physico-chemical parameters. Their mean values were presented in the Table 7.1. The physical parameter like appearance of sample–1 is turbid and sample–2 is colourless and clear. The sample 2 falls under normal or maximum permissible limit. Turbidity of sample 1 is 105.33 and sample 2 is 0.278. The turbidity of sample 2 falls under normal or maximum permissible limit. Total dissolved solids of sample 1 are 181.7 and sample-2 is 594. The TDS of sample 2 falls under normal or maximum acceptable limit. The electrical conductivity of sample 1 is 251.5 and sample 2 is 860. Electrical conductivity of sample 2 falls under normal or maximum acceptable limit.

The chemical parameters like pH, Alkalinity, Total hardness, Calcium, Magnesium, Sodium, Potassium, Iron, Manganese, Free ammonia, Nitrite, Nitrate, Chloride, Fluoride, Sulphate, Phosphate, Tidy's test, Silica and Biological oxygen demand of sample–1 is 7.2,63.2, 54.3, 12.1, 5.7, 30.97, 4.9, 6.89, Nil, 1.69, Nil, 7, 33.18, 0.609, 15.9, 0.134, 1.370, 1.222 and 5.8 and the sample-2 is 7.12, 173.3, 291.9, 70.65, 28.85, 58.3, 6.89, 0.147, Nil, 0.305, Nil, 14.45, 140.9, 0.387, 50.2, 0.295, 0.0211, 48.23 and 2.7 respectively.

The pH, Magnesium, Sodium, Potassium, Iron, Manganese, Free ammonia, Nitrite, Nitrate, Fluoride, Sulphate, Tidy's test of both the samples fall under normal or maximum acceptable limit. Alkalinity, Total hardness, Calcium, Chloride and Biological oxygen demand of the sample-1 are below the normal or maximum acceptable limit. Alkalinity, Total hardness, Calcium, Chloride and Biological oxygen demand of sample 2 fall under normal or maximum acceptable limit.

Discussion

Availability of water and its quality is very important for the healthy living of all organisms on earth. Ecosystem's health in this biosphere is considerably influenced not only by the quantity of water but also by the quality of water. In this present investigation physical and chemical parameters of the lake water and Guru Nanak College underground water was analyzed and based on the results, the physicochemical parameters of both the water samples falls under the normal or maximum acceptable limits.

Table 7.1: Physico-chemical Parameters of the Water Samples of Vattambakkam Lake

Sl.No.	Parameters	Normal/Maximum Acceptable Limits	Sample 1	Sample 2
1.	Appearance	Colourless	C&C	C&C
2.	Odour	Unobjectionable	None	None
3.	Turbidity (NTU)	5–10	105.33±0.700	0.278±0.434
4.	Total dissolved solids(mg/L)	500–1500	181.7±0.715	594±1.256
5.	Electrical conductivity (micro mho/Cm)	1000–4770	251.5±0.703	860±0.558
6.	pH	6.5–9.2	7.2±0.0237	7.12±0.007
7.	Alkalinity (as $CaCO_3$) (mg/L)	200–600	63.2±0.213	173.3±0.831
8.	Total hardness (as $CaCO_3$) (mg/L)	300–600	54.3±0.309	291.9±0.525
9.	Calcium (as Ca) (mg/L)	200–1000	12.1±0.191	70.65±0.433
10.	Magnesium (as Mg) (mg/L)	30	5.7±0.253	28.85±0.477
11.	Sodium (as Na) (mg/L)	150	30.97±0.472	58.3±0.633
12.	Potassium (as K) (mg/L)	12	4.9±0.152	6.89±0.143
13.	Iron (as Fe) (mg/L)	1	6.89±0.142	0.147±0.066
14.	Manganese (as Mn) (mg/L)	0.05–0.17	Nil	Nil
15.	Free Ammonia (as NH_3) (mg/L)	3	1.696±0.140	0.305±0.012
16.	Nitrite (as NO_2) (mg/L)	1	Nil	Nil
17.	Nitrate (as NO_3) (mg/L)	45	7±0.244	14.25±0.967
18.	Chloride (as Cl) (mg/L)	250–1000	33.18±0.858	140.9±1.784
19.	Fluoride (as F) (mg/L)	1–1.5	0.609±0.024	0.387±0.129
20.	Sulphate (as SO_4) (mg/L)	200–450	15.9±0.348	50.2±0.785
21.	Phosphate (as PO_4) (mg/L)	0.02–0.11	0.134±0.005	0.295±0.13
22.	Tidy's test (as O) (mg/L)	0.024	1.370±0.014	0.0211±0.001
23.	Silica	–	1.222±0.006	48.23±0.278
24.	BOD (mg/L)	3.5	5.8±0.133	2.7±0.152

C&C: Colourless and Clear.

Each values are expressed as mean±SEM of ten samples.

The characteristics of water depending on water resources are as follows according to Anon (1996). The characteristics of water available of human use depends on its source. Normally rain water is found to recharge various sources like surface water, sub surface water in the river beds and ground water.

The surface water may be running stream or river or it may be a pond or lake. The characteristics of a surface water when not polluted will be typical in the sense that certain parameter like colour, turbidity, silica, ferric, iron, and bacterial load will be present in high level and certain other parameters like Electrical Conductivity (EC), Total Dissolved Solids (TDS), Total Hardness (TH), Calcium (Ca), Magnesium (Mg), Sodium (Na), Potassium (K), Nitrate (No_3), Chloride (cl), Fluoride (Fl) and Sulphate (SO_4) will be present in low levels. The surface water found in hills and mountain will have very low TDS (10-50 mg /L) while it will be more in the surface water of the plains (100-500 mg /L).

The infiltration wells and galleries fall under the category of sub surface water sources. They are dug in the river beds, the level of most natural parameters in these sources will be low and in the acceptable range.

Dug wells, shallow borewells, deep borewells and tube wells fall under ground water sources category. Wide variation will be noticed in the quality of water is found from place to place. Generally the water quality parameters in ground water sources colour, turbidity, silica, ferric, iron, and bacterial load will be present in low level while Electrical Conductivity (EC), Total Dissolved Solids (TDS), Total Hardness (TH), Calcium (Ca), Magnesium (Mg), Sodium (Na), Potassium (K), Nitrate (No_3), Chloride (Cl), Fluoride (Fl) and Sulphate (SO_4) will be present in high levels. Now-a-days the several ground water sources have high level of nitrate (50-100 mg/ L) due to application of high amount of fertilizers.

When water travels through soil depending upon the nature of soil, the level of some parameters present in trace level may exceed the permissible level. Fluoride is such one parameter. The level of pesticides, detergents, oil and grease etc. are also increasing day by day in natural water. Apart from this, due to pollution from industries, toxic materials like cyanide, arsenic, phenols etc., are also finding access to natural water. The Biological Oxygen Demand (BOD) is contributed by several industries producing organic substances.

The results obtained were compared with standard values of APHA (1995) and found that both the samples of Vattambakkam Lake and Guru Nanak College campus underground water are within permissible limits for drinking and domestic uses and are free from pollution.

References

Anon., 1996. A book on Training programme for new entrant junior water analysts conducted by RDT center, Tamil Nadu water supply and drainage board, Chennai.

A.P.H.A. 1995. Standard methods for the examination of water and waste water, American Public Health Association.

Meitei, N.S., Bhargava, V. And Patil., P.M. 2004. Water quality of Purna river in Purna town area of Maharashtra state. *J. Aqua Biol.* 19 (2): 77-78.

Trivedy, R.K. and Chandrasekhar, T.R.1999. Sediment characteristics of fresh water bodies of Mangalore, Karnataka. *J. Eco. Biol.* 11 (1): 59-64.

Chapter 8

Water Quality Index (WQI) of Shetty Lake and Hadhinaru Lake of Mysore District, Karnataka, India

B.M. Sudeep, S. Srikantswamy and S.P. Hosmani

Introduction

In recent years, there has been considerable research on nutrient dynamics, pollution control (and abatement) of lake water bodies. The value and importance of freshwater resources necessitates that they be well managed ecologically for meeting water quality standards. Along this line work continues in the field on: identifying general properties, ecosystem characteristics, human impacts and utility of empirical models for resource management (Cooke *et al.*, 1993; Li and Yang, 1995; Moss *et al.*, 1996; Havens *et al.*, 2001; Havens and Schelske, 2001; Timchenko *et al.*, 2000). The National Sanitation Foundation (NSF) developed WQI which was based on a few physico-chemical parameters. This index was suitable for testing drinking water supplies (Princy *et al.*, 2001). According to Zaheeruddin and Khurshid (1998), Manish and Pawar (1998), Hari Haran (2002) rapid industrial growth, urbanization and agricultural activities are the main factors for pollution of water. Chinmoy and Raziuddin (2002) and Agarkar (1998) have reported that waste water also changes the physical parameters like colour, smell, taste and such waters can create health problems in consuming populations. Freeda *et al.* (2002), Sharma *et al.* (2001), and Guruparasad (2003) have emphasized the importance of water quality. Beena *et al.* (2001), and Naik and Purohit (2001) state that physical and chemical characteristics of water can adversely effect ecological balance in many ways. Ramesh *et al.* (2001) is of the view that storm water harvesting and removal of limiting nitrates from urban water run off are among the promising solutions in regulating the entery of nutrients into water. Agarkar (1998), Alam and Mubeen (1998), Baber and Kaplay (1999) noted TDS and total hardness values of some water were above standard permissible limits indicating sources of waste water and sanitary disposal in the near by surroundings.

WQI indicates the quality of water for any intended use. It is defined as a rating reflecting the composite influences of different water quality parameters on the overall quality of water. The indices are among the most effective ways to communicate the information on water quality trends to the general public or the policy makers and in water quality management. The objectives of this study was to assess the status of water quality in Shetty lake and Hadhinaru lake of Mysore district, Karnataka state, India for a period of one year (January 2005–December 2005).

Materials and Methods

Water samples were collected from January 2005–December 2005; once in the first week of the month between 8am to 10am from the study area.

Study Site

Shetty lake is located at 12º 14' 14" N and 76º 39' 37" E at an attitude of 701.6 MSL It is situated 4kms away from Mysore city, (Figure 8.1). The water in it is used for agriculture and fishing with an

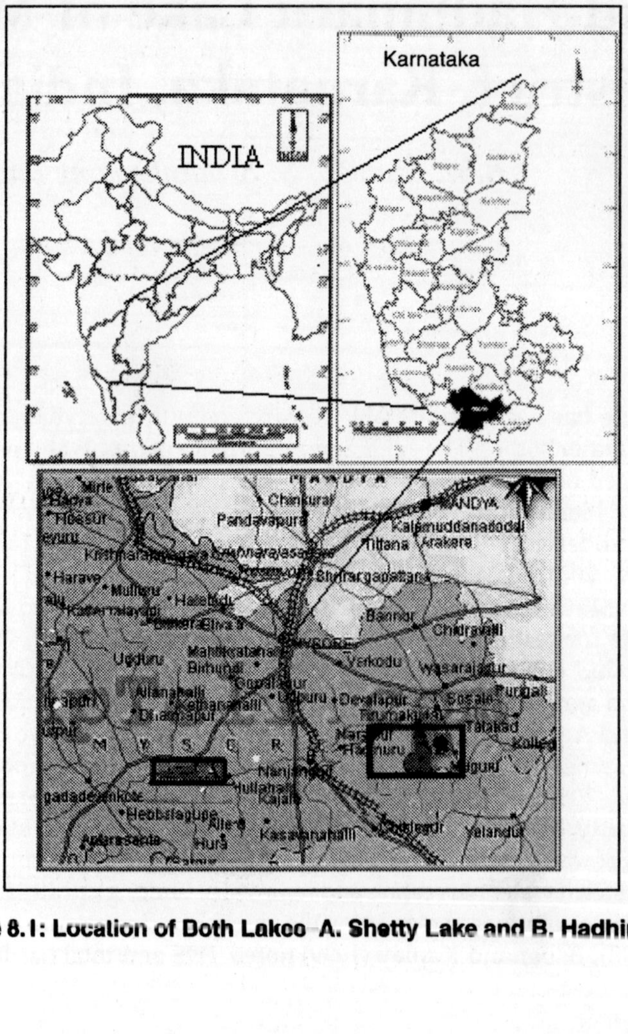

Figure 8.1: Location of Both Lakes A. Shetty Lake and B. Hadhinaru Lake

independent catchments area of 1.79sq.km; spread over an area of 4.8hecters with a live capacity of 9Mcft. The maximum depth of the lake when full is seven meters and has become polluted due to continuous discharge of concentrated sewage, agricultural runoff as well as contaminated water from the neighbouring Mandakalli lake which is also polluted. It is a perennial lake. Hadhinaru lake is located at 12° 02' 02" North longitudes 76° 41' 38" East latitude at an altitude of 653.35 MSL. It is situated 33kms away from Mysore city (Plate 8.2). It has an independent catchments area of 8.57sq.km, with water spread area of 10.10 hectares having a live capacity of 54.43 MCFT. The maximum depth of the lake when full is five meters. The lake bund has a top width of 10, front slope 1.1, near slope 2.1, and free board 3 meters.

Sample Collection

The physical and chemical characterization of each station was accomplished from water and sediment samples collected at the margins. Water samples were collected in a clean acid washed polyethylene two liter bottle.

The following parameters were measured: pH was measured immediately on board. Sample bottles were kept in the dark at low temperature and further processed in the laboratory. Dissolved oxygen was determined by Winkler method. BOD was determined from the decrease in oxygen concentration after 5- days of incubation in the dark. Total alkalinity and total hardness were measured titrimetrically. TDS was measured by evaporation method. Samples were dried at 105°C for at least 1 h. Sulphate was measured by gravimetric method with drying residue. Chloride was measured by argentometric precipitation method. All the above mentioned chemical parameters were estimated according to the standard procedures given in APHA, 1995.

Results and Discussion

Calculation of WQI of selection parameters has great importance. Since selection of many parameters may widen the WQI and importance of various parameters depends on the intended use of water. Among them eight physico-chemical parameters such as pH, TDS, total alkalinity, total hardness, chloride, sulphate, DO and BOD are used to calculate WQI. The calculation of WQI was done using Weighted Arithmetic Index method (Brown *et al.*, 1972) in the following steps.

Calculation of Water Quality Index

Calculation of the Subindex or Quality Rating (Q_N)

Let there be n water quality parameters and quality rating or subindex (q_n) corresponding to the nth parameter is a number reflecting the relative value of this parameter in the polluted water with respect to its standard permissible value. The q_n is calculated using the following expression:

$$q_n = 100\,[(Vn - Vi0)/s_n - V_n)]$$

where

q_n: Quality rating for the nth water quality parameter.

V_n: Estimated value of the nth parameter at a given sampling station.

s_n: Standard permissible value of the nth parameter.

V_{i0}: Ideal value for nth parameter in pure water.

All the ideal values (V_{i0}) are taken as zero for the drinking water except for pH= 7.0 and DO = 14.6 mg/l.

Calculation of Quality Rating for pH

For pH ideal value is 7.0 (for neutral water and 8.5 for polluted water). Therefore, quality rating for pH is calculated from the following reaction:

$$q\,pH = 100\ [(V\,pH - 7.0)/8.5 - 7.0)]$$

where,

V pH: Observed value of pH.

Calculation of Quality Rating for DO

The ideal value (V_{Do}) for DO is 14.6 mg/l and standard permissible value for drinking water is 5 mg/l. therefore, quality rating is calculated from following relation.

$$qDO = 100\ [(V\,DO - 14.6)/5 - 14.6)]$$

where,

V_{DO}: Measured value of DO

Calculation of Unit Weight (W_N)

The unit weight (W_N) for various water quality parameters is inversely proportional to the recommended standards for the corresponding parameters.

$$W_N = K/\ Sn$$

where,

Wn: Unit weight of nth parameters

Sn: Standard value for nth parameter

K: Constant for proportionality

WQI

WQI is calculated from the following equation

$$(WQI) = \sum_{n=1}^{n} q_n w_n \Big/ \text{Ó}\,Vi_0$$

The rating of WQI was calculated from 8 parameters as given in the Table 8.1 and Table 8.5. The calculated values of both the lakes are presents in Table 8.1-8.4 for Shetty lake and Tables 8.6–8.8 for Hadhinaru lake. Drinking water recommending agencies standards are presented in Tables 8.9 and 8.10.

The monthly WQI of Shetty lake indicates that the value of 284 represents unsuitable for drinking purpose. Similarly the values of WQI for Hadhinaru lake were below 100 during most of the months except during March (153.7) ar.d April (139.00). During March and April the water reached the state of unsuitability for drinking, but during other months also the water had a very poor quality. The monthly variations are represented in Figure 8.2.

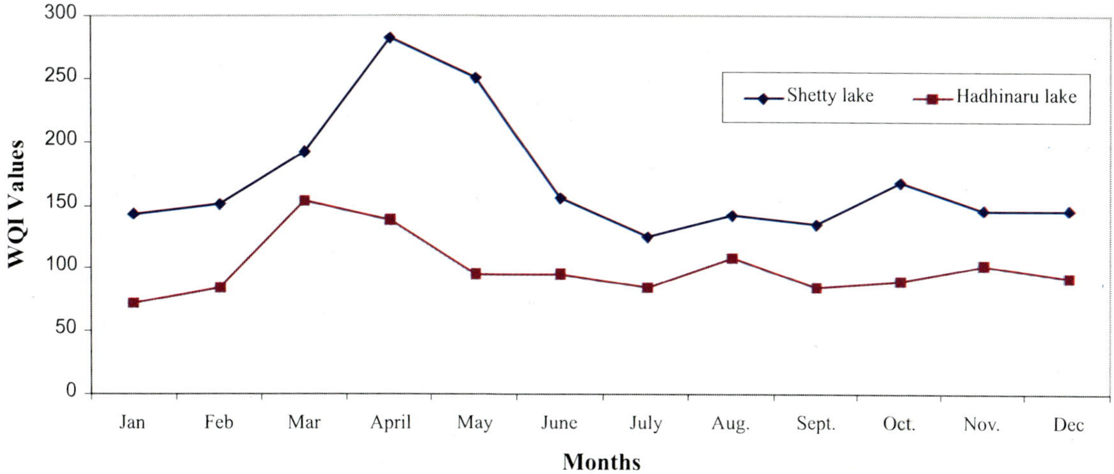

Figure 8.2: Monthly WQI Values of Shetty Lake and Hadhinaru Lake

Proper sanitation and protection of seepage of sewage water entering into the lake must be done. The chemical characteristics of water are of special interest to the users because the usefulness of the supply is frequently dependent upon the mineral content of the water. The suppliers and the users should have an idea of the limits of the chemical substances which are specified in the drinking water standards and such of the waters that are of poor quality should be excluded from supply systems. Both the lake waters Shetty lake and Hadhinaru lake are unsuitable for drinking water supply as per the rating of the WQI and require proper treatment.

Table 8.1: Water Quality Index of Shetty Lake

Parameter	*Jan*	*Feb*	*Mar*	*April*	*May*	*June*	*July*	*Aug.*	*Sept.*	*Oct.*	*Nov.*	*Dec*
pH	8.6	8.8	9	8.0	0.0	8.7	8.6	8.7	8.6	8.5	8.6	8.7
DO (mg/L)	3.4	3.6	3.8	3.8	4.6	5.2	10.2	17.3	6.4	5.8	5.2	7.1
BOD (mg/L)	9.5	10.2	16	26.4	25	12	11	14	10.2	14	10.8	12
Total alkalinity (mg/L)	484	470	260	270	240	440	307	280	300	680	520	240
Total hardness (mg/L)	145	175	208	252	250	258	270	280	256	244	250	206
Chlorides (mg/L)	98	168	116	282	280	146	152	104	128	110	106	96
Sulphate (mg/L)	13	30	36	38	36	34	29	28	34	10	11	34
TDS (mg/L)	476	625	720	648	622	650	694	780	510	496	506	479

Figure 8.3: Southern View of Shetty Lake

Figure 8.4: Southern View of Hadhinaru Lake

Table 8.2: Water Quality Rating (q_n) of Shetty Lake

Parameter	*Jan*	*Feb*	*Mar*	*April*	*May*	*June*	*July*	*Aug.*	*Sept.*	*Oct.*	*Nov.*	*Dec*
pH	106.6	120	133.3	190	106.6	113	106.6	113.3	106.6	100	106.6	113.3
DO (mg/L)	116.6	114.5	112.5	113	104.1	97.9	45.83	28.12	85.4	91.66	97.9	78.12
BOD (mg/L)	190	204	320	528	500	240	220	280	204	280	216	240
Total alkalinity (mg/L)	403.3	391.6	216.6	225	200	366.6	255.8	233.3	250	566.6	433.3	200
Total hardness (mg/L)	48.3	58.3	69.3	84	83.3	86	90	93	85.3	81.3	83.3	68.6
Chlorides (mg/L)	39.2	67.2	46.4	112.8	112	58.4	60.4	41.6	51.2	44	42.4	38.4
Sulphate (mg/L)	5.2	12	14.4	15.2	14.4	13.6	11.6	11.2	13.6	4	4.4	13.6
TDS (mg/L)	95.2	125	144	129.6	124.4	130	138.8	156	102	99.2	101.2	95.8

Table 8.3: Calculated Subindex Values ($Q_n w_n$) Values of Different Parameters at Shetty Lake

Parameter	*Jan*	*Feb*	*Mar*	*April*	*May*	*June*	*July*	*Aug.*	*Sept.*	*Oct.*	*Nov.*	*Dec*
pH	23.23	26.16	29.05	41.42	23.23	24.6	23.2	24.6	23.23	21.8	23.23	24.69
DO (mg/L)	43.2	42.46	41.7	41.9	38.6	36.3	16.29	10.82	31.67	33.99	36.3	28.9
BOD (mg/L)	70.4	75.6	118.6	195.8	185.4	89.01	81.5	103.8	75.6	103.8	80.1	89.01
Total alkalinity (mg/L)	6.23	6.05	3.34	3.47	3.09	5.66	3.95	3.6	3.86	8.75	6.69	3.09
Total hardness (mg/L)	0.298	0.36	0.428	0.51	0.514	0.53	0.55	0.57	0.52	0.5	0.51	0.42
Chlorides (mg/L)	0.29	0.49	0.344	0.83	0.83	0.43	0.45	0.3	0.37	0.32	0.31	0.284
Sulphate (mg/L)	0.03	0.08	0.1	0.11	0.1	0.1	0.086	0.083	0.1	0.02	0.032	0.1
TDS (mg/L)	0.352	0.46	0.53	0.47	0.46	0.48	0.51	0.57	0.37	0.36	0.37	0.35
Monthly WQI level	143.96	151.5	194.0	284.3	252.1	157.1	126.24	143.8	135.65	168.92	147.4	146.7

Table 8.4: Over All Water Quality Index of Shetty Lake

Parameter	$\Sigma q_n w_n$	Σw_n
pH	308.4	2.616
DO (mg/L)	402.13	0.044
BOD (mg/L)	1269	0.019
Total alkalinity (mg/L)	57.78	0.74
Total hardness (mg/L)	5.71	0.089
Chlorides (mg/L)	5.248	0.089
Sulphate (mg/L)	0.941	4.45
TDS (mg/L)	5.282	4.45
	2054.491 (Σ Qi Wi)	12.66 (Σ Wi)

Water Quality Index (WQI) = $\Sigma q_n w_n / \Sigma w_n$

= 2054.491 (Σ Qi Wi)/12.497 = 162.28 (unsuitable for drinking – Shetty lake).

Table 8.5: Water Quality Index of Hadhinaru Lake

Parameter	Jan	Feb	Mar	April	May	June	July	Aug.	Sept.	Oct.	Nov.	Dec
pH	8.2	8.6	8. 3	8.6	8.2	8	7.9	7. 9	7. 8	8	8. 5	8. 5
DO (mg/L)	7.2	8.11	4.05	5.27	8.12	10.22	10.2	12.19	9.32	9.73	10.14	10.54
BOD (mg/L)	3	4.5	12	10	6.8	8	7	11	6.2	6	8.2	7.98
Total alkalinity (mg/L)	359	252	320	330	240	400	185	440	300	340	234	250
Total hardness (mg/L)	84	116	204	280	140	200	240	200	120	132	152	156
Chlorides (mg/L)	71	82.3	85.2	106.5	67.4	85.2	71.7	68.16	105	51.12	71	71.4
Sulphate (mg/L)	8	11	10	24	8	10	6	4	11	7	8	8

Table 8.6: Water Quality Rating (q_n) of Hadhinaru Lake

Parameter	Jan	Feb	Mar	April	May	June	July	Aug.	Sept.	Oct.	Nov.	Dec
pH	80	106	86	106	80	66.6	60	60	53.3	66.6	100	100
DO (mg/L)	77.08	67.6	109.8	97.1	65.9	45.6	45.6	25.1	55	50.72	46.45	42.29
BOD (mg/L)	60	90	240	200	136	160	140	220	124	120	164	159.6
Total alkalinity (mg/L)	299	208	266	275	200	333	154.1	366.6	250	283.3	195	208.3
Total hardness (mg/L)	28	38.6	68	93.3	46.6	66.6	80	66.6	40	44	50.6	52
Chlorides (mg/L)	28.4	32.94	34.08	42.6	26.94	34.08	28.68	27.26	42	20.44	28.4	28.56
Sulphate (mg/L)	3.2	4.4	4	9.6	3.2	4	2.4	1.6	4.4	2.8	3.2	3.2

Table 8.7: Calculated Subindex Values ($q_n w_n$) of Different Parameters at Hadhinaru Lake

Parameter	Jan	Feb	Mar	April	May	June	July	Aug.	Sept.	Oct.	Nov.	Dec
pH	17.45	23.12	18.76	23.16	17.45	14.39	13.09	11.62	14.39	21.81	21.81	14.39
DO (mg/L)	28.58	25.07	40.72	36.01	24.44	16.9	16.9	9.309	20.39	18.81	17.22	15.68
BOD (mg/L)	22.25	33.3	89	74.17	50.44	59.34	51.92	81.59	45.99	44.5	60.82	59.19
Total alkalinity (mg/L)	4.61	3.21	4.1	4.24	3.09	5.14	2.38	5.66	3.86	4.37	3.01	3.21
Total hardness (mg/L)	0.173	0.238	0.42	0.57	0.28	0.41	0.49	0.41	0.24	0.27	0.31	0.32
Chlorides (mg/L)	0.21	0.24	0.25	0.31	0.19	0.25	0.21	0.2	0.31	0.151	0.21	0.211
Sulphate (mg/L)	0.023	0.032	0.029	0.071	0.023	0.029	0.017	0.011	0.032	0.02	0.023	0.023
TDS (mg/L)	0.37	0.43	0.46	0.54	0.341	0.57	0.64	0.574	0.652	0.66	0.55	0.48
Monthly WQI level	73.63	85.60	153.7	139.00	96.20	96.98	85.60	109.32	85.864	90.59	103.39	93.504

Table 8.8: Over all Water Quality Index of Hadhinaru Lake

Parameter	$\Sigma q_n w_n$	Σw_n
pH	211.4	2.61
DO (mg/L)	270	0.044
BOD (mg/L)	672.5	0.185
Total alkalinity (mg/L)	46.88	0.74
Total hardness (mg/L)	4.131	0.089
Chlorides (mg/L)	2.742	0.089
Sulphate (mg/L)	0.333	4.45
TDS (mg/L)	6.267	4.45
	1214(Σ Qi Wi)	12.66(Σ Wi)

Water quality index (WQI) = $\Sigma q_n w_n / \Sigma w_n$
= 1213.378/12.497 = 95.89 (poor water quality)

Table 8.9: Drinking Water Standards Recommending Agencies and Unit Weights

Sl.No.	*Parameters*	*Standards (sn)*	*Recommending Agency*	*Unit Weights (w_n)*
1.	pH	7.0-8.5	ICMR	0.218176
2.	DO (mg/L)	500	WHO	0.003708
3.	BOD (mg/L)	120	ICMR	0.01545
4.	Total alkalinity (mg/L)	300	ICMR	0.0061816
5.	Total hardness (mg/L)	250	ICMR	0.0074179
6.	Chlorides (mg/L)	250	WHO	0.0074179
7.	Sulphate (mg/L)	5.0	ICMR	0.037089
8.	TDS (mg/L)	5.0	ICMR	0.37089

(All values except pH are in (mg/L)).

Table 8.10: Status of Water Qualities Based on WQI (Mishra and Patel)

WQI	*Status*
0–25	Excellent
26–50	Good
51–75	Poor
76–100	Very poor
100–above	Unsuitable for drinking

References

Alam, M. and Mabeen, 1998. Ground Water quality in Jamia Nagar and adjoining areas Orientation. J. *Chem.* 14 (2): 333–335.

Agarkar, S.V., 1998. Assessment of Water Quality of Sakegoan reservoir Maharashtra. *Asian J. Chem.*10 (4): 997–998.

American Public Health Association. 1995. Standard Methods for Examination of Water and Wastewater, 16th edition. Washington.

Babar, M.D. and R.D.C Kaplay, 1999. Ground water quality around Pingaland nala in Prabhani district, Maharashtra, India. *Eco. Envt. Conserv.* 5 (2):141-143.

Beena, T., Absaham and T.M.S. Begum, 2001. Comparison of water quality of wells and bore wells of Mootha kunnam, Malinkara area with USPHS Standards. Indian *J. Environ and Ecoplan*. 5 (3) 593: 596

Brown, R.M., N.J. Mecleiland, R.A. Deinmger and M.F. O'Conner, 1972. A water quality index crassing the physiological barrier (Jenkis, S.H.ed). Proc. Int. Conf, on *Water Poll. Res.* Jerusalem. 6: 787–797.

Cooke, G.D., Welch, E.B., Peterson, S.A. and Newroth, P.R. 1993. Restoration and management of lakes and reservoirs. 2nd ed. Boca Raton, FL: Lewis Publishers; 1993.

Chinmoy, C. and M. Raziuddin, 2002. Determination of Water Quality Index (WQI) of a degraded river in Asanol industrial area (West Bengue). *Nat. Envt and Poll. Tech.* I (2): 181–189.

Freeda, D. G., R. S. Geetha and J. Ebancizor, 2002. The drinking water quality characteristics of five rural places in and around Thiltagud, Tamilnadu. *Poll. Res.* 211 (2): 215–221.

Havens, K.E. and Schelske, C.L., 2001. The importance of considering biological processes when setting Total Maximum Daily Loads (TMDL) for phosphorus in shallow lakes and reservoirs. *Environ. Pollut.*, 113: 1–9.

Guruprasad, 2003. Evaluation of water quality in Tadepallimandal of Guntur district A.P. *Nat. Envt. and Poll. Tech.* 3: 273–276.

Li, W. and Yang, Q., 1995. Wetland utilization in Lake Taihu for fish farming and improvement of lake water quality. *Ecol. Eng.*, 5: 107–121.

Moss, B., Stansfield, J., Irvine, K., Perrows, M. and Phillips, G., 1996. Progressive restoration of a shallow lake: a 12-year experiment in isolation, sediment removal and biomanipulation. *J. Appl. Ecol.*, 33: 71–86.

Manish, C. V. and K.T. Pawan, 1998. Assessment of drinking water quality of an industrial township of South Bihar. *Jour of Envt and Poll.*, 5 (1): 17–21.

Naik S.K and K.M. Purohit, 2001. Status of Water quality at Bondamui Rourkel Industrial Complex Part II: Metallic parameters Indian. *J. Environ. and Ecoplan.* 5 (1): 115–118

Princy, J. M., I.S. Lydia and A. Rajendran, 2001. Formulation of new Water Quality Index WQI–3. *Ecotoxicol. Environ. Monit.* 11(2): 91- 99.

Timchenko, V., Oksiyuk, O. and Gore, J., 2000. A model for ecosystem state and water quality management in the Dnieper River delta. *Ecol. Eng.*, 16: 119–25.

Hari Haran, A.V.L.N.S.H., 2002. Evaluation of drinking water quality at Jolarpet villege of Vishakapatanam district, Andhra Pradesh. *Nat. Envt. and Poll Tech.* 1 (4): 407–410.

Ramesh, H.S., S. Mahesh, M. Mahadeva Swamy and R.S. Abijit, 2001. Assessment of nutrients in urban storm water runoff in Mysore city. *Poll. Res.* 20 (2): 215–218.

Sharma, K.C., I. Hassan, J. Hussani and K.G. Ojha, 2001. Ground water quality of an industrial town Bhilwara, Rajasthan. *J. of Envt and Poll* 8: 110–114.

Zaheeruddin and S. Khurshid, 1998. Dispersion of heavy metals and its toxic effects on water quality in the parts of Faridabad. Haryana. *Ecotoxical Environ. Monit.* 8 (1): 9–14.

Chapter 9

Preliminary Observations on Physico-chemical Characteristics of Water and Soil of Domestic Sewage Water and It's Suitability for Aquaculture

S.S. Dhane, S.J. Meshram and P.E. Shinagre

Introduction

India is rich in water resources, being endowed with a network of rivers and vast alluvial basins to hold ground water. However, it's natural variability in the existence and availability are of much greater importance. Out of the total annual precipitation of about 400 m ha M (Million hectare meters) in India, 188 m ha M flows off as runoff of this, it is estimated that only about 70-75 m ha M is usable after provision for requisite storage, without major inter-basin transfer. Depending upon population projections the projected demand for water in Cu.ton. is given in Table 9.1.

Table 9.1: Projected Demand for Water (Cu. ton)

Purpose	*Year, 2000*	*Year, 2025*
Domestic use	033	052
Irrigation	630	770
Energy	027	071
Industry	030	120
Others	030	037

Source: FAO Report, 2000.

As population going on increasing day-by-day water requirement is going increasing. Water utilization for irrigation is 90 per cent and only 5 per cent is utilized for community water supply.As water resources going on decreasing it is need to utilize domestic sewage water for irrigation and aquaculture.

Domestic Sewage Water

The Water, which discharged from the residential areas into the eco-system after domestic use, is considered as wastewater. It contains food waste, water from lavatories and baths etc. A typical sewage is relatively dilute 90-99 per cent water, grayish brown, odiferous, fluid-semi fluid containing organic and inorganic solids in dissolved and suspended forms (Welch, 1993). It has been estimated to contain ammonia, Cal, Nitrogen 3.92 to 11.36 mg/l, Nitrite–nitrogen 0.12–0.32 mg/l, Phosphate–Phosphorus 1.2–9.18 mg/l, Calcium 40.6–75.0 mg/l, Magnesium 16.0–30.0 mg/l, Potassium 3.0–9.8 mg/l and in India total volume of available sewage has been quantified to be about 3,637 MLD, which contain about 90 tonnes N, 32 tonnes P_2O_5 and 55 tonnes K_2O (Naskar and Shaha, 1988, Patri, 1997). The use of this water was first started in Mumbai in 1897, in Delhi from 1913 and in Pune from 1916 for growing various crops. In West Bengal near about 4000ha area was used for the culture of fishes in domestic sewage water. It is reported that around 20 per cent fishes consumed by the Calcutta city people is supplied through sewage culture (WHO, 1989).

It is therefore, desirable to optimally use the available fresh water resources and also was convicted to use domestic waste water for aquaculture. Direct use of domestic sewage water is harmful for culture of fishes as it contains high amount of CO_2, BOD, harmful gases and some amount of pollutant. Therefore to use the water for aquaculture it is very essential to treat the water either with biological filter or mechanical filter, so that water gets clear which having near about 3-4 ppm oxygen.

As this domestic sewage water contains high amount of organic and inorganic material the soils in the fishpond gets affected. Therefore it is essential to study the soil and water parameter of domestic water for aquaculture point of view. In view of this domestic wastewater coming from nearby residential colonies through Panvel Municipal nala is analyzed for aquaculture point of view. Important physico–chemical properties of soil and water which deciding usability of the said domestic water are presented and discussed.

Materials and Methods

The domestic waste water coming from nearby residential colonies through Panvel Municipal nala was diverted in the farm nala crude separation of sundry material such as plastic bags etc. was done with the help of iron net fixed in nala itself after that it was collected in sedimentation tank and finally it had gets filter through course gravel filter. This filtered water gets stored in pond and this water can be useful for aquaculture point of view.

The Chemical analysis such as pH, EC, salt content, osmotic pressure, calcium, magnesium, sodium, Nitrate, free sulphides and saline ammonia content, BOD, COD content of said water have been determined by adopting standard methods of analysis (APHA, 1989).

The Physico-chemical analysis of soils such as sand, silt, and clay percentage of soil, pH, organic carbon percentage, total organic matter have been analyzed by adopting standard procedures (Jackson, 1973). The water was also got analyzed for heavy metals like Copper, Mercury, Zinc, Cadmium and Nickel from the Maharashtra Pollution Control Board Laboratory, Navi Mumbai.

Results

Sewage water analyses for aquaculture are presented in Table 9.2 and Physico–chemical analysis of soil parameter are given in Table 9.3.

All the parameters in raw sewage water were found to be beyond the safe limit or favourable range required for aquaculture. But after sedimentation and coarse gravel filteration water gets purified which showed water in favourable range for aquaculture. Most of the parameters of fish pond were either in favourable range or just nearby it. Where as in soil parameters not much difference found in sand, silt and clay percentage of soil. But increase in organic carbon percentage, total organic matter, total nitrogen were found after filling the domestic sewage water in fish pond.

Table 9.2: Sewage Water Analysis for Aquaculture

Sl.No.	Parameter	At intake	After Sedimentation	After Course Gravel Filter	Fish Pond	Favourable Range
1.	pH	6.9	7.1	7.2	7.4	7.0–8.5
2.	Alkalinity (mg/l)	296	256	144	168	>50
3.	Hardness (mg/l)	258	240	230	180	50–180
4.	Chloride (mg/l)	70	60	51	166	<50
5.	Salinity (ppt)	Nil	Nil	Nil	Nil	<0.50
6.	Dissolved oxygen (mg/l)	1.8	4.4	4.6	4.2	5–10
7.	Dissolved free carbon dioxide (mg/l)	16	8	6	Nil	<3
8.	Total solids mg/l	431	357	298	241	<500
9.	Total suspended solids (mg/l)	400	45	30	117	<30
10.	Biochemical oxygen demand (mg/l)	310	73	25	8	<15
11.	Chemical oxygen Demand (mg/l)	640	170	90	14	<50
12.	Ammonia–Nitrogen (NH_4+–N: mg/l)	35	10	1.5	0.25	0–1.0
13.	Nitrite–Nitrogen (NO_2–N: mg/l)	2.7	0.50	0.30	0.24	0–0.5
14.	Nitragae–nitrogen (NO_3–N: mg/l)	1.12	5.14	5.35	0.60	0.1–3.0
15.	Phosphate (PO_4–P: mg/l)	2.50	2.0	0.50	0.01	0.05–0.4

Among the heavy metals copper, Lead, Cadmium, Nickel were found to be absent. While Zinc and Mercury were recorded as 0.06 and 0.05 mg/lit, respectively in sewage water (Table 9.4). The bio–assay studies indicated that sample is non-toxic.

Discussion

From the above results of Physico–Chemical parameters of soil and water it is clear that, if city sewage water is treated with sedimentation and coarse gravel filteration system, water quality gets improved and found to be in the favourable range required for aquaculture. Further concentrations of heavy metal was found to be within safe limits. (WHO, 1989). Mukherjee *et al.* (2002) explained the sewage fed fish culture practice in West Bengal. In this practice, entire Kolkata Municipal Sewage is collected in the settling tank and after settling, it was distributed to the aquaculture system. Ghosh (1983) reported that nutrients and COD values are reduced to safe limits before discharge of water to open area using sewage–fed–fish culture. He also indicated that treatment and reuse of sewage water in aquaculture is economically feasible method to reduce the pollution level. Being preliminary studies, the use of aquatic weed as biological medium for removal of pollutants is not worked out. Howerver, studies carried out by many research workers (Das and Jana, 1999; Duffer, 1982 and Tripathi and Shukla, 1991) have revealed the possibility of using plants as biological fillers for water purification.

Table 9.3: Soil Parameters in Fish Pond Before and After Filling of Pond with Domestic Sewage Water

Sl.No.	*Parameter*	*Surface*	*Middle*	*Bottom*	*Pond Aoil After Filling Water*	*Favourable Rane*
1.	Sand (per cent)	30	32	37	32	30
2.	Silt (per cent)	28	27	31	35	30
3.	Clay (per cent)	42	41	32	33	40
4.	pH	7.0	7.0	7.2	7.2	6.0–8.0
5.	Water retention capacity (per cent)	50.3	50.5	46.8	46.0	>40
6.	Organic carbon (per cent)	1.33	1.27	1.20	1.40	70.50
7.	Total organic matter (per cent)	2.29	2.18	2.06	2.41	>0.86
8.	Total nitrogen (mg/100gm of soil sample)	133	127	120	140	>50

Table 9.4: Analysis of Heavy Metals in Domestic Sewage Water

Sl.No.	*Heavy Metals*	*Available in Water (mg/l)*	*Safe Limits (mg/l)*
1.	Copper	Nil	25
2.	Lead	Nil	100
3.	Cadmium	Nil	10
4.	Nickle	Nil	10
5.	Zinc	0.06	100
6.	Mercury	0.05	0.1

Conclusions

The results obtained under preliminary studies on physico–chemical properties of sewage water before and after treatment showed that the water quality can be improved to such an extent to carry out the aquaculture practices. As available water source is decreasing, it is an opportunity to utilize domestic sewage water for aquaculture. Further, reuse of this waste water will be economically feasible method to minimize treatment cost and obtains maximum benefits through aquaculture by using nutrients contained is the wastewater. There is further scope to improve the water quality by using biological filtrations systems to make best use of domestic sewage water for aquaculture.

References

APHA 1989. Standard Methods for the examination of water and wastewater. Am. Public Health Association Washington, DC, 17th edn., 1452 pp.

Das, S. and Jana, B.B. 1999. *Eichhernia Crassipes* induced cadmium depuration in tilapia. *O. mossambicus* (Peters). *J. Aqua. trop*. 14 (3) 201-207.

Duffer, W.R. 1982. Assessement of aquaculture for reclamation of waste water. In Water Reuse (Ed.) E.J. Midle Brooks. Ann Arbor Sc. Ann Arbor, Mich; 349-369.

Ghosh, D., 1998. Empowering rural communities for waste water treatment and reuse, ten, lessons from replicating. Calcutta Wetland experience. Meenbarta, special issue on wet land, 16 June, 1998,department of Fisheries, Government of West Bengal, Calcutta, 42-46 pp.

Jackson, M.L. 1973. *Soil Chemical Analysis*. Prentic Hall, New Delhi, pp, 498.

Mukherjee, M., U. Nath, S.K.Kashem and M. Chattopadhyoy, 2002, Peri- urban Aquaculture and its environmental status: Kolkata Experience. *Fishing chimes*: 22 (1): 9 to 14.

Naskar, K.R. and S.K. Sahi, 1988. Utilization of domestic wastes for integrated farming in the tropic. *J. Aqua. Trop*., 3: 37-41.

Patri Priyambada, 1999. Community liquid waste application through aquaculture. *Yojana*, Feb: 21-22.

Tripathi, B.D. and Shukla, S.C. 1991. Biological treatment of wastewater by selected aquatic plants. *Environ. Pollut*. 69-78.

Welch, E.B. 1993. Ecological effects of wastewater applied limnology and pollutant effects. Champanan and Hall: 425 P.

WHO, 1989. Health guideline for use of wastewater in agriculture and aquaculture, WHO Tech. Rep. Ser. No. 778 WHO, Geneva.

Chapter 10

Assessment of Microbial Load Along with Limnological Analysis in Flood Affected Areas of Thar Desert of Rajasthan, India

Mamta Rawat, Sumit Dookia and G.R. Jakher

Introduction

India offers immense climatic diversity and topographical varieties and desert form the backdrop of many a legend in India. Thar or Great Indian Desert is an arid region 800 km long and 400 km wide, in North West of India and East of Pakistan, between the Indus and Sutlej River on the west and the Aravalis Range on the east. At present, the Thar Desert of Rajasthan is spreaded over a 28,600 km^2 which is 62 per cent of the total area of the State (Tyagi and Yadav, 1996). Largely a desolate region of shifting sand dunes, broken rocks, and scrub vegetation, receives an annual average rainfall of less than 25 cm that too scanty and erratic in nature, with prolonged drought and sudden rainfall with all its variations.

Although large parts of Rajasthan are drought-prone, flash floods (overtopping defined ephemeral rivers/streams and spreading in flood plains over a short duration) are not uncommon in the state. Flash floods are common hydrological features in arid western America, Central Australia and arid northwest India (Hoyt and Langbein, 1955; Mabbutt, 1977; Ramaswamy, 1975; Rakhecha *et al*., 1979). Drainage basins of high relief, a large percentage of surficial bedrock sparse vegetation and shallow soils are susceptible to flash flooding when cyclonic storms or well-marked low-pressure areas produce intense and localized rainfall. By definition, a flash flood is short-lived, with sharp-peaked hydrographs and with floodwaters rising and falling almost equally rapidly with sudden appearance and disappearance in the headwater streams (Ward, 1978). There were episodes of flash floods in some parts of western Rajasthan, way back in 1973, 1975 and 1976. Whereas during 1979, large parts of the state witnessed flash floods due to a very heavy downpour of more than 500 mm in 3-5 days, which cut off the arid region from rest of the country for several days (Sharma and Vangani, 1982).

The monsoon (rain of July-October) of year 2006, proved to be a deluge in spite of showers of blessing as usual. A sudden precipitation of more than 600 mm rain within 2 days, created a big dilemma in the form of heavy flood in this sandy landscape and very soon the desert turned into the sea. Some villages like Kawas, Malwa and Bhadaka in Barmer district turned into the biggest mortuary in the Desert. The most affected village, Kawas, is still submerged under 4-7 feet of water, after 11 months of the rain in August last year. A thick layer of gypsum in Kawas, the worst hit area in the district, was preventing floodwaters from percolating, as there are thick layers of clays-bentonite and fullers earth (Tertiary) and gypsite (Quaternary) at a depth of about 1-1.5-foot in this area (Laul and Laul, 2007). The possibilities of being non-productive land, which comes under the submergence, also can't be overlooked. As the bentonite and gypsite bed in the under ground will increase the salinity of the soil, along with the ground water, through capillary action.

The continue water logged area of these villages, with hard under ground of gypsite and bentonite, neither supports the percolation nor having much help in the drainage, make this large amount of water into unusable source. Continues water logging in this desert makes chaos of ecological changes and that suddenly develop interest of various workers in this part. In this order, a case study was undertaken to find out the water quality with respect to its limno-microbial status along with influence of flash flood on health impacts. There were three specific aims of this study:

1. To identify the major health–impacting microbiological and physico-chemical contaminants of flood water.
2. To determine the severity of contamination by selected contaminants including indicator and pathogens, and identify the potential risk factors for contamination.
3. To correlate the disease outbreak with microbial community.

Materials and Methods

The village Kawas (25° 52′ 20.75″ N and 71° 32′ 57.41″ E), Bhakada (25° 53′ 00.15″ N and 71° 32′ 57.41″ E) and Malwa (25° 52′ 25.21″ N and 71° 32′ 55.12″ E) most effected one, is situated in Barmer district of western Rajasthan, whereas, village Malwa is situated about 70 km away towards west of Barmer city and Bhadaka on northwest side of city. The complete submerged area were surveyed and sampled from October 2006 to April 2007, when the rainwater receded and a temporary pathway was being formed by local authorities. Though the water level was still about 3 feet to 4 feets in Kawas and almost a huge lake of more then 30 sq km was formed, the water samples were taken from different places.

Water Sampling

Samples were preferably drawn from the every possible site, *viz.*, the distance from every 5 to 10 of flooded area in order to get the proper data or status of water quality. So, overall, 7 sampling stations or sites were sampled for testing water quality from these 3 villages in Barmer district.

Physico-chemical Sampling and Analysis

Samples were analyzed for temperature, pH, conductivity, dissolved oxygen, free carbon dioxide, alkalinity, salinity, and nutrients (nitrate, phosphate and silicates) respectively. All parameters were measured using standard methods of APHA (1998).

Microbiological Sampling and Analysis

Microbiological samples were taken directly into the individual sterile 250 ml water bottles and transported to the laboratory in a thermostat chamber. Samples were then usually analysed on the same day as the distance of sampling sites and research laboratory was not too much. The presence-absence test for indicator organisms including Total Coliform (TC), Faecal Coliform (FC) and Feacal Streptococcus (FS) were assessed as per standard methods of APHA (1998). Analysis was also performed for a range of pathogenic bacterial species including *Salmonella* spp., *Staphylococcus*, *Pseudomonas* and *E. coli* EHEC serotype 0157:H7.

Isolation and Identification of *Salmonella*

Appropriate amount of sample was inoculated into Selenite F broth for enrichment and incubated for 18±hrs. at 37 ° C. Following the incubation of Selenite Fenrichment broth, a loopful was streaked onto Bismuth Sulphite (BIS) and Xylose-Lysine-Deoxycholate (XLD) agar plates in duplicates and incubated aerobically at 37° C for 24 hr. One to three colonies suspected to be Salmonella (small red colonies on XLD agar and black-brown colonies on BIS agar) were selected, purified by streaking onto nutrient agar plates and characterized bio chemically using Kligler Iron Agar (KIA) and Urease tests. Himedia strips were used for further species confirmation of genus Salmonella.

Isolation and Identification of *E. coli* EHEC Serotype 0157:H7

About 100ml of sample and 50 ml of triple strength lauryl tryptose broth was incubated at 37 c for 24 hours. It was then streaked on Sorbitol McConkey (SMAC) agar containing potassium tellurite.The sorbitol negative strains (SMAC) were tested for enzyme B-d- glucuronisade using MUG(4-methylumbelliferyl–b-d glucuronide) as substrate.A colony of the test isolate was applied to MUG impregnated paper, moistened with a drop of saline and incubated at 37 for 20 min. The absence of enzyme was indicated by the lack of fluorescence when the filter paper was examined under U.V. light. The MUG negative colonies were further serotype using 0157 and H7 antiserum.

Isolation and Identification of *Staphyloccous* spp.

The suspected gram positive cocci were further tested by slide test. A drop of saline and suspected colonies was emulsified. An immediate formation of granular precipitate of white clumps within 15-20 seconds interprets for positive strains of Staphylococcus, *i.e., S. aureus*. Non-coagulating strains were reported as coagulase negative staphylococcus.

Isolation and Identification of *Pseudomonas* sp.

Gram-negative bacilli colonies were brought in contact with soaked strips of tetra methyl derivative of p-phenyl diamine.A instant production of blue colour on the contact area of impregnated strips confirm the pseudomonas species. As the blue coloration takes place due to cytochrome oxidase activity.

Results and Discussion

Limnological Aspects

The limnogical quality of stagnant floodwater revealed that water was quite soft in nature and almost all limnological parameters lied within safe range according to World Health Organization (WHO, 1996) criteria (Table 10.1). The hydrogen ion concentration (pH) of the stagnant rainwater was observed within alkaline range which meets the standard limits of WHO, 1996 (6.5-8.5) and the same results were also shown by Yaziz *et al*. (1989). Rich amount of dissolved oxygen (Kawas = 6.55mg/lt,

Malwa = 7.02 mg/lt, Bhadaka = 6.85 mg/lt) favored the possibility of survival of aquatic life like fish fries and aquatic birds on them. The presence and survival of fish fries in 3 to 10 feet depth of rain-hoarded water depicts the sufficient amount of dissolve oxygen in it. However Johnston *et al.* (2003), experienced situation of deoxygenating resulting of fish kills in sulfidic back swamps.

Table 10.1: Limnological Parameters from Flooded Area of Villages

Parameters	*Average Values±SD*			*WHO Guidelines*
	Kawas (n=7)	*Malwa (n=7)*	*Bhadaka (n=7)*	
Temperature	24.8	22.6	23.0	–
pH	6.8	6.4	7.2	6.5–8.5
Dissolved oxygen (mg/l)	6.55	7.02	6.85	–
Carbonate alkalinity (mg/l)	28	56	34	–
Bicarbonate alkalinity (mg/l)	107	209	380	–
Chloride(mg/l)	57	65	44	250
Calcium hardness (mg/ll)	239	220	390	75
Magnesium hardness (mg/l)	270	106	185	50
Total hardness (mg/l)	1420	1260	1350	300
Nitrate (mg/l)	15	8	12	20
Phosphate (mg/l)	1.4	0.8	0.56	1
Silicate (mg/l)	5.75	3.05	1.85	250

In terms of the dissolved substances floods are often viewed as dilution phenomenon in which low conductivity rainwater dilutes groundwater or base flow that are nearer chemical equilibrium with substrata and thus rich in dissolved salts and therefore low concentrations during floods were observed (Sharma and Vangani, 1982). The high values of total dissolved solids are mainly due to the lithology of the area.

The values for total hardness were observed somewhat high (1350 mg/l), probably due to the deposition of gypsum beneath the soil strata. The entire nutrients studied (nitrate, phosphate and silicate) were found within safe limits. Dillon and Jeffrey (2005) concluded that nutrients (ammonium, nitrate and phosphate) values elevated during storm relative to rainwater. In the present context the chlorine levels were not very high.The average value of chloride level was 57,65 and 44 mg/l at respective villages (Table 10.1). Johnston *et al.* (2003) also reported high level of chemical composition from two sulfidic swamps into the Clarence River estuary after the flooding. Whereas the hygienic investigations made by Fenske *et al.* (1997) in the water of Odra lagoon along with the beaches of Usedom, resulted in good quality values, therefore swimmers or bathers were not at risk, even during the flood.

Biological Aspects

Though the floodwater was studied by various workers world over, but on limno-microbial aspects of flash flood of desert region is not studied yet, most probably due to the reason that such phenomenon of flash floods are not frequent. Floods are not only accompanied by disaster of life and wealth but they also leave the after effects in the form of disease caused by the stagnant water, submerged dead bodies of animals livestock and wild and human beings. The scenario of post flood assessment of these villages reveals that the prime threat was likely to come from water borne diseases like jaundice, typhoid, cholera, etc. which eventually caused dysentery, nausea, loose stools, abdominal tenderness and occasional fever which all lead to severe dehydration and, in extreme cases, death. A bacteriological study of stagnant floodwater of Kavas village reveals that apart from indicator bacteria, many pathogenic bacteria were also found. The indicator bacterial species, *Escherichia coli* and *Streptococcus feacalis* were found in considerable number, along with pathogenic bacteria namely, *Salmonella enterica and Staphylococcus aureus*. The colonies of *E.coli* were observed maximum in village Bhakada (83,000 colonies/100 ml) followed by Kawas (79,800 colonies /100ml) and Malwa villages (32,800 colonies /100 ml).The FC/FS ratio observed out at three respective villages (1.99, 2.04 and 3.77) depicted that the contamination was from both human and non-human sources. Another bacterium that is a common coloniser of human skin and mucosa and a causative organism of illnesses such as skin and wound infections, urinary tract infections, pneumonia and bactercaemia (blood stream infection) besides food poisoning was also found in abundance at the samples of three villages (Table 10.2). Apart from *S. aureus*, a group of positove cocci, coagulase negative Staphylococcus, was found maximum in Kawas village (4,450 colonies/100 ml), followed by Malwa (2930 colonies/100 ml) and Bhadaka (1530 colonies/100 ml) village. High numbers of *Staphylococcus* could have lead to skin and wound infections, but the control measures taken by health experts at appropriate time did not let this happen.The strains of *Salmonella* are the main cause of gastrointestinal illness and typhoid fever. *Salmonella enterica* was observed in the stagnant rain water of all three villages, so the occurrence of gastrointestinal illness and typhoid fever as per reported by local news papers can be directly contributed to it. Pathogenic *E.coli*, EHEC serotype 0157:H7, is a causative agent of acute diarrhoea,was only present at Kawas village (35 colonies/100ml).

Table 10.2: Bacteriological Parameters from Flooded Areas of Villages

Parametres	*Average Colony Counts/100 ml*			*WHO Guidelines*
	Kawas	*Malwa*	*Bhadaka*	
Total Bacterial Colonies (TBC)	1,81,000	1,00,670	1,56.200	Nil
Total Coliform Colonies(TCC)	1,25,500	89,780	1,02,700	Nil
Feacal Coliform Colonies(FCC)	79,800	67,000	83,000	Nil
Feacal *Streptococcus* Colonies(FSC)	40,040	32,800	22,010	Nil
Pseudomonas spp	8,960	7600	10400	Nil
Staphylococcus aureus	3,500	4200	7890	Nil
Coagulase negative *Staphylococcus*	4,450	2930	1530	Nil
Salmonella enterica	980	1720	560	Nil
E.coli EHEC serotype 0157:H7	35	–	–	Nil
FC/FS	1.99	2.04	3.77	Nil

The presence of various pathogenic strains in the flooded areas in not uncommon. *Salmonella typhi* contaminated sewage was aspirated into the main water distribution system after Hurricane Claudette's flooding, evidenced by high water marks on the walls (Morgan *et al.*, 2005). In a large study undertaken in Indonesia in 1992–1993, flooding was identified as a significant risk factor for diarrhoeal illnesses caused by *Salmonella enterica* serotype Paratyphi A (paratyphoid fever) (Vollard *et al.*, 2004). Other nonspecific infections such as conjunctivitis and ear, nose, and throat infections also increase after flood (WHO, 2004).

In the flood areas, where the population does not have access to clean water and sanitation,there is potential for increased feacal oral transmision of disease. There was a marginal increase in gastroenteritis and other water borne diseases cases, but later on the situation were kept under control. The sudden presence of large numbers cattle carcasses in the disaster affected raised concerns of outbreak of diseases.Apart from reports from UNICEF, Rajasthan, of the marginal increase in gastroenteritis and other water borne diseases through the reports of UNICEF, Rajasthan. Health experts of Rajasthan also warned that there could be an outbreak of other water-borne diseases like typhoid, jaundice and malaria in the flooded areas if the administration doesn't clear out the stagnant water (Anonymous, n.d.). However, the situation did not turned in the very severe way, except mosquito borne diseases like Malaria and Chickengunia and few cases of typhoid fever and no doubt gastritis cases.

A systematic review of recently published literature found limited epidemiological evidence about the health effects from flooding disasters (Ahern *et al.*, 2005). Notably, there were fewer studies from developing countries, where the disease burden is likely to be higher (WHO, 2004). The review found that diseases transmitted by the faecal–oral route were the main flood-related health impact, like nonspecific diseases, *i.e.*, diarrhoea, cholera, dysentery, and typhoid (Gerrity and Flynn,1997). For example, diarrhoea increased by a factor of two to four after flooding in Mozambique during 2000 (Kondo *et al.*, 2002). Although increased incidence of diarrhoea disease in affected populations is not usually associated with increased mortality, there have been some exceptions. Diarrhoeal disease outbreaks can occur after drinking water has been contaminated and have been reported after flooding and related displacement from various flood-affected areas. According to WHO reports (WHO, 2006) up to 40,000 cases of acute diarrhoeal disease have been reported in Ethiopia alone during the flooding, including 403 deaths in December 2006. It is believed that drinking and household water sources have been contaminated since toilet pits and shallow wells for drinking water in the affected areas have collapsed or have been buried under the floodwaters (Anonymous, 2007). An outbreak of diarrhoeal disease after flooding in Bangladesh in 2004 involved >17,000 cases; *Vibrio cholerae* (O1 Ogawa and O1 Inaba) and enterotoxigenic *Escherichia coli* were isolated (Qadri, 2005). Flooding in West Bengal in 1998 was followed by an outbreak of diarrhoea, suspected to be cholera (O1 Ogawa), which resulted in 16,590 cases and 276 deaths (case-fatality ratio 1.7 per cent) (Sur *et al.*, 2000) and floods in Mozambique in January-March 2000 led to an increase in the incidence of diarrhoea (Kondo *et al.*, 2002). The risk for diarrhoeal disease outbreaks following natural disasters is higher in developing countries than in industrialized countries (Noji, 1997 and Ahern *et al.*, 2005). In the United States, diarrhoeal illness was noted after Hurricanes Allison (Waring *et al.*, 2002) and Katrina (CDC, 2005a, 2005b), and norovirus, *Salmonella*, and toxigenic and nontoxigenic *V. cholerae* were confirmed among Katrina evacuees. After natural disasters, the media, health professionals and relief workers often say that dead bodied of victims can cause epidemics of diseases such as cholera (Goyet, 2000). However, victims of natural disasters are unlikely to harbour pathogenic organisms such as cholera, which can cause epidemics. For the public, the risk of infectious disease from dead bodies after natural disasters is negligible. However, the risk of infection can be greatly reduced by following basic hygiene precautions (Morgan, 2004).

The relation between flooding and vector-borne disease is complex. Many important infections are transmitted by mosquitoes, which breed in stagnant water. Flooding may also create a large number of breeding sites for mosquito-borne diseases such as malaria, and there have been a great increase in number of mosquitoes through various villages of Barmer district. This increase can be particularly important when populations are displaced. Similar situation have been reported after the Mozambique floods of 2000, the number of malaria cases within the displaced population increased by a factor of 1.5 to 2 times previous levels (Kondo *et al.*, 2002). Also malaria and dengue fever passed by mosquitoes were also reported in high numbers. A number of cases of malarial diseases as after effect of floods from various parts of world are mentioned by Ahern *et al.* (2005).

A systematic and comprehensive evaluation made after this flash flood concludes that though a high numbers of indicator and pathogenic bacteria were observed in the stagnant flood water, however the incidences of that rate of mortality was not observed in the area. A rapid action controlled the situation. Also, Healing *et al.* (1995) concluded in their studies that common gastrointestinal organisms do not survive long in the environment and present little risk of infection where the body has been decaying for sometime or has been in the water.

Social Impact on the Communities

The Thar Desert in Rajasthan, which basks in sunshine for around 320 days a year, was transformed into a trough of death and destruction when it experienced unprecedented flash floods in August 2006. The locals get heavily affected, not only economically but also lost hundreds of human life as well as livestock. Economically, floods are a leading cause of losses from natural events. Flood only is not a single disaster event, but also creates a cumulative loss from related small and medium sized events. The flash flood of August 2006 of the Barmer district alone, worst hit area impacted, of the 104 human deaths in Barmer, 46 were children below 15, more than half of the 300 reported missing are also children. Of the children who survived, over 600 in Barmer are reportedly suffering from acute gastroenteritis and other water-borne diseases (Anonymous, 2006). It also make a heavy toll on other properties of local inhabitants including, 93,526 houses/huts destroyed, nearly 1,07,000 cattle perished, about 55,867 hectare land rendered unproductive due to sand-casting, 3,180 wells buried or damaged, and 119 wells were even completely buried. Apart from this, extensive damage was caused to various infrastructures such as rail and road (107 km length), transportation, electric installation, telecommunication, water supply schemes, mining and industrial units and socio-economic amenities.

Flash Floods and their Potential

The frequent occurrence of severe flash floods in drought prone Western Rajasthan suggest that an opportunity exists to convert a part of the floodwater into groundwater recharge by constructing a large number of groundwater recharge structures, which are non-existent now. The water of flash floods if managed and utilized for rejuvenating the depleted aquifers in the state, the issue of water scarcity during droughts can be alleviated to a large extent. Excess rain in the drought-vulnerable region with sandy terrain can be diverted to groundwater aquifers and surface storage structures. Floodwater farming has been practiced in the desert areas of Arizona and northwest New Mexico for at least the last 1000 years (Zaunderer and Hutchinson, 1988). In the *"Khadin"* system of India, floodwater is impounded behind earth bunds, and crops then planted into the residual moisture when the water infiltrates.

Conclusions

From the above analysis it is apparent that unprecedented heavy and widespread monsoon rainfall of single night duration was greatly intensified by saturated antecedent gypsum and bentonite deposited beneath soil and a larger area of exposed sand dunes did not let the water of flash flood drain off in the kavas village. Neither, the flood subsided rapidly and inundation continues for a period of more than seven months in the depressions of gypsum deposited soil. The after effects of flood were even devastating as there were frequent episodes of acute gastrointestinal illness. Weaker sections of the society were affected more due to the unprecedented flash flood, which also retarded the progress of the future planning. Appropriate and timely intervention can significantly reduce the risk of mortality and morbidity from infectious diseases after flooding disasters. In the short-term, preventing the diseases spread via the faecal–oral route is the most important public health intervention. This relies on three measures: provision of clean water, suitable sanitation, and hygiene promotion (The Sphere Project, 2004). As there is no natural drainage system in the desert, which normally faces droughts, the sustained relief operations will take a longer duration.

Acknowledgements

This case study was carried out after the flood in the Thar Desert, to quantify the aftermath of flood in severely affected areas of Barmer District of Rajasthan. The first author wishes to thanks Department of Science and Technology, for providing financial assistance through OYS-Fast Track Scheme (SR/FT/L-128/2004 dated 2.6.2005).

References

Ahern, M., Kovats, R. S., Wilkinson, P., Few, R., Matthies, F. 2005. Global health impacts of floods: epidemiologic evidence. *Epidemiological Review*, 27: 36–46.

American Public Health Association (APHA). 1998. Standard methods for the examination of water and wastewater. 20th ed. American Public Health Association, Washington, DC.

Anonymous.2006. *Floods that changed Rajasthan's landscape, rhythm.* "http://www.nerve.in/news:25350019790", assessed on 16.10.2006.

Anonymous.2007.Rajasthan's:Flood-hit Barmer still suffering. "http://news.indiainfo.com/2007/03/05/rajasthan-farrners.html",assessed on 5.3.2007.

Anonymous (not dated). *Children, the most vulnerable.* "http://www.unicef.org/india/media_2284.htm", assessed on 12. 6.2007.

Centers for Disease Control and Prevention (CDC).2005a. Two cases of toxigenic *Vibrio cholerae* O1 infection after Hurricanes Katrina and Rita—Louisiana. *Morbidity and Mortality Weekly Response*, 55:31–2.

Centers for Disease Control and Prevention (CDC).2005b. Norovirus outbreak among evacuees from hurricane Katrina—Houston, Texas, September. *Morbidity and Mortality Weekly Response*, 54:1016–8.

Dillon, K. S. and Jeffrey, P. C.2005. Nutrient transformations between rainfall and stormwater runoff in an urbanized coastal environment: Sarasota Bay, Florida. *Limnological Oceanography*, 50: 62–69.

Gerrity E, Flynn B.1997. Communicable diseases and disease control after disasters. In: Noji E, editor. The public health consequences of disasters Oxford: Oxford University Press.

Goyet C, de Ville de.2000. Stop propagating disaster myths. *Lancet* 356: 762–764.

Healing, T. D., Hoffman, P. N. and Young, S. E.1995.The infectious hazards of human cadavers. *Comun. Dis. Rep. CDR Rev.* 5(5):R61-8.

Hoyt, W.G. and Langbein, W. D.1955.*Floods*. Princeton University Press, Princeton, New Jersey, USA. 496 pp.

Johnston, S. G., Slavich, P. G., Sullivan, L. A. and Hirst, P. 2003. Artificial drainage of floodwaters from sulfidic backswamps: effects on deoxygenation in an Australian estuary. Marine and Freshwater Research, 54: 781-795.

Kondo, H., Seo, N., Yasuda, T., Hasizume, M., Koido, Y., Ninomiya, N. and Yamamoto, Y.2002. Post-flood–infectious diseases in Mozambique. *Prehospital Disaster Medicine*, 17:126–133.

Laul, V. P. and Laul, V. 2007. Floods in Barmer and other Tertiary terrains in Rajasthan. *Current Science*, 92: 568.

Mabbutt, J. A. 1977. *Desert Landforms*. Australian National University Press, Canberra. 151-159 pp.

Morgan, O.2004. Infectious disease risk of dead bodies following natural disasters. *Revista Panamericana de Salud Pública*, 15: 307–312.

Morgan, O., Ahern, M. and Cairncross, S. (2005). Revisiting the Tsunami: Health Consequences of Flooding. *PLoS Medicine*, 2(6): 184-491-493.

Noji, E.1997.*Public health consequences of disasters* (ed.). New York: Oxford University Press, UK.

Qadri, F., Khan, A.I., Faruque, A. S. G., Begum, Y. A., Chowdhury, F., and Nair, G. B. 2005. Enterotoxigenic *Escherichia coli* and *Vibrio cholerae* diarrhoea, Bangladesh. Emerging Infectious Disease, 11: 1104-1107.

Rakhecha, P. R., Kulkarni, A. K., Mandai, B. N., Dhar, O. N. and Chakraborty, A. K. 1979. Heavy rainfall distribution over Rajasthan. In: *Water Resources Management in Rajasthan* (Proc. Jaipur Symp.), 12-27 pp.

Ramaswamy, C.1975. Floods in the desert. In: *Environmental Analysis of the Thar Desert* (ed. by R.K.Gupta and I.Prakash), 134-137. English Book Depot, Dehrudun, India.

Sharma, K. D. and Vangani, N. S. (1982). Flash flood of July 1979 in the Luni basin–a rare event in the Indian desert. *Hydrological Sciences–Journal–des Sciences Hydrologiques*, 27, 385-397.

Sur, D., Dutta, P., Nair, G., and Bhattacharya, S.2000. Severe cholera outbreak following floods in a northern district of West Bengal. *Indian Journal of Medical Research*, 112:178–182.

The Sphere Project.2004. Humanitarian charter and minimum standards in emergency response. Oxford. Oxfam Publication, UK.

Tyagi, B.K. and. Yadav, S.P.1996. Malariological and sociological significance of '*tanka*' and '*beri*' in the Thar Desert, Western Rajasthan, India. *Journal of Arid Environments*, 33: 497–501

Vollard, A. M., Ali, S., van Asten, H. A., Widjaja, S., Visser, L. G., and Surjadi, C. 2004. Risk factors for typhoid and paratyphoid fever in Jakarta, Indonesia. *Journal of American Medical Association*, 91: 2607–15.

Ward, R.1978. *Floods. A Geographical Perspective*, 13-27. Macmillan Press, UK.

Waring, S. C., Reynolds, K. M., D'Souza, G., Arafat, R. R.2002.Rapid assessment of household needs in the Houston area after Tropical Storm Allison. *Disaster Management Response*, 3–9.

WHO.1996. Guidelines for drinking water quality. Vol.2, Recommendations. World Health Organization, Geneva.

World Health Organization.2004. Flooding and communicable diseases fact sheet. http://www.who.int/hac/techguidance/ems/FloodingandCommunicableDiseasesfactsheet.pdf. Accessed on 10.9.2007.

Yaziz, M. I., Gunting, H., Sapari, N., and Ghazali, A. W.1989. Variations in rainwater quality from roof catchments. *Water Research*, 23: 761–765 pp.

Zaunderer, J and Hutchinson, C. F.1988. *A review of water harvesting techniques of the Arid Southwestern US and North Mexico.* Working paper for the World Bank's Sub-Sahara Water Harvesting Study.

Chapter 11

Diversity of Fish Fauna in Nagathibelagulu Pond, Shimoga, Karnataka

H.M. Ashashree, M. Venkateshwarlu and H.M. Renuka Swamy

Introduction

India is home for more than 10 per cent of global fish biodiversity with 2,200 species of fish and shellfish, 1, 440 in marine, 103 in brackish water, 544 in fresh water and 73 in cold water. Over 7 million people are directly engaged in different fishery and aquaculture activities of country (Ayyappan, 2007).

India is endowed with a vast expanse of open inland water. There are about 3153,366 hectare reservoirs, 202,212 hectare lakes, 2,200,000 hectare ponds, besides 29,000 km length of rivers (Sugunan, 1999). Even Karnataka has large number of freshwater bodies which includes 33,612 minor tanks, 2,767 major tanks, 18 reservoirs, and 3,000miles of rivers, innumerable irrigation channels, wells and temple ponds (Fish seed committee report, Government of India., 1998).

In India fish culture is practiced in ponds, because pond diversity represented by a number of aquatic plants, and animals including vegetation, planktons, weeds and various bottom dwelling forms. So this type of biodiversity is absolutely essential for stability of different ecosystem for proper functioning of their food chain. Fishes is a highly perishable commodity constituted by 60-80 per cent and with lesser amount of fat. Hence a detailed study of fish fauna of an area forms the prerequisite for understanding any fish culture program on a scientific basis keeping the view the present study was undertaken.

Materials and Methods

Sampling of fish has been made for every fortnight days throughout study period. Collection of fish was made directly from the fishermen's during the time of fishing. Two types of fish nets were used (1) Gill net (2) Cast net. The collected fishes were preserving in 4 per cent formalin. Later on in the laboratory they were identified on the basis of their morphological characters with the help of Jayaram (1981), Talwar and Jhingran (1991).

Description of the Study Area

The Nagathibelagalu pond is associated with Sri. Nanjundeshwara temple. Temple is center of worship for Hindus. The Nagathibelagalu pond is about 8 km away from Bhadravathi town and it belongs to Shimoga district of Karnataka State. The pond is perennial one and acquires a total area of about 25 acres and possess an average depth of 10-15 feet. The pond receives rain water and channel water as main water sources. The pond is used for washing and bathing by devotees and local community. Water quality is an important criteria for fish habitat hence the physico-chemical characters of water from the pond was analyzed (Table 11.1)

Table 11.1: Water Quality of Nagatibelagalu Pond

Sl.No	*Parameters*	*Value*
1.	Temperature (°C)	25
2.	pH	8
4.	Dissolved CO_2 (mg/ltr)	26
5.	Total alkalinity (mg/ltr)	125
6.	Chloride (mg/ltr)	31.24
7.	Total hardness (mg/ltr)	60.6
8.	BOD (mg/ltr)	8.07

Results and Discussion

The fish fauna is an important aspect of fishery potential of a water body. It is observed that the distribution of fish species is quite variable because of geographical and geological conditions. In the present study 15 fish species have been identified. (Table 11.2) This fish species belongs to 5 orders. The order siluriformes and order perciformes were dominant with 5 species each followed by order cypriniformes (3 species), and Cyprinodontiformes and Beloniformes with one species each.

The diversity of fishes mainly depends upon the biotic and abiotic factors and types of the ecosystem (lentic and lotic habitat).The physico-chemical properties of water was analyzed following the standard procedure APHA (1998). The study revealed that the pond is suitable for fish culture (Table 11.1). The pond is natural one, but here more amount of culturable fishes are found and local people consume almost all the species.

Table 11.2: List of Fish Species in Nagatibelagulu Pond

Sl.No	Order	Family	Species
1.	Cypriniformes	Rasbornidae	*Rasbora daniconius, Rasbora rasbora*
		Balitoridae	*Lepidocephalus thermalis*
2.	Perciformes	Channidae	*Channa striatus, Channa punctatus, Channa marulius*
		Cichilidae	*Oreochromis mossambica*
		Ambassisidae	*Ambassis kopli bleeker*
3.	Siluriformes	Claridae	*Clarias batrachus*
		Siluridae	*Ompak bimaculatus*
		Bagridae	*Mystus cavasius, Mystus seenghala, Aorichthys aor*
4.	Cypriniodontiformes	Poecilidae	*Poecillia reticulatus*
5.	Beloniformes	Horichthyinae	*Horaichthys setnai*

References

A.P.H.A. 1980. Standard methods for the examination of water arid waste water. American public health association. Washington D.C.

Arvind, N., Shukla, Anil Pandey and Sharad Shrivastava, 2003. Observation on fish biodiversity variation in Gambhir dam, Ujjain (M.P). India.*Aquacult*.4 (1): 15-21.

Ayyappan, S. 2007. Growth through inland aquaculture, National Academy of Agricultural Sciences Congress held at Tamil Nadu Agricultural University, Coimbatore, India.

Calestousjuma, 1998. The CBD and the biological diversity of Inland waters. Jacques, suite 300, Montreal, Quebec, Canada.

Jayaram, K.C. 1981. The freshwater fishes of India, Pakistan, Bangladesh, Burma and Sri Lanka a hand book, SZI, Calcutta, India.

Talwar, P.K and Jhingharan, A. G. 1991. Inland fishes of India and adjacent countries; volumes I and II- Oxford and IBH pub. House, New Delhi, India.

Sugunan V.V. 1999. Fisheries management of small water bodies in seven countries of Africa and Asia- A review. *Fishing Chimes*. 19: 7-10.

Chapter 12

Culture Potential of Clams Along the Coast of India

Milind M. Girkar, Sachin B. Satam and Prakash Shingare

Introduction

Edible bivalves forms cheap but important source of nutrition and employment generation, especially to the coastal peoples. Along the Coast of India vast areas including muddy bays, rocky inshore areas, estuaries and backwaters are known to support the molluscan resources comprising of clams, oysters and mussels forming the regular fishery along the coast.

The venerid clam, *Paphia malbarica* commonly known as false clam contributes 80-90 per cent of the total frozen export from India (Appukuttan *et.al.*, 1999). The clam meat, rich in protein, is the most preferred mollusc by the local peoples for regular consumption. Of late, the clam has gained popularity as supplementary food in semi-intensive prawn farming during the entire culture period including nursery phase (Kripa and Gopakumar, 1996). Besides, the clam shells, which are rich in calcium carbonate, also form the raw material in the manufacture of paper, chemicals, polyfibre, cement and lime based industries (Kripa and Joseph, 1993).

Biology of Important Clam Species

Like oysters and mussels, the clams are filter feeders, feeds mainly on phytoplankton and detritus. A larva feeds on microalgae *Isochrysis galbana*. Sexes are separate. Clams which burrow into the substratum with the help of well developed foot.

Selection of Site for Clam Culture

Clams are cultured on the bottom and therefore following points should be considered while site selected. (Narsimham and Laxmilatha, 1996).

1. Availability of natural seed.

Table 12.1: Resources and Distribution

Sl.No.	State	Dominant Species	Estuary
1.	Maharashtra	*Meretrix meretrix, K. opima, P. malbarica*	Kalbadevi, Mirya, Bhatye and Vijaydurga estuary
2.	Goa	*Meretrix casta, Villorita cyprinoids*	Mandavi estuary
3.	Karnataka	*Meretrix meretrix, Paphia malbarica, M. casta, K. opima, Villorita cyprinoids*	Kundapur, Mullcy, Gurpur and Udyavara, Netravati, Swarna, Uppunda and Sita.
4.	Kerala	*Paphia malbarica, Villorita cyprinoids*	Koduvally, Azzikal, Chitturi, Vembanad lake and all estuaries.
5.	Tamilnadu	*Meretrix casta*	Vellar and Pulicat
6.	A. P.	*Meretrix meretrix, M. casta, K. opima, Paphia malbarica, Anadara granosa*	Bhimunipatnam, Kakinada bay, Godavari estuary
7.	Orissa	*Meretrix sp.*	Chilka lake and Sonapur back waters
8.	Andaman	*Tridaena maxima, Tridaena crocea, Tridaena squmosa*	Andaman coast line

Source: Handbook on Aquafarming Molluscs (P.S.B.R. James and K.A. Narsimham, C.M.F.R.I., Kochi, Kerala.

Table 12.2: Key for an Identification of Important Genera

Sl.No.	Species	Important Identification Characters
1.	*Paphia malbarica*	1. Shell is concentric grooved through the impressions. 2. Shell length is 1/3 times long than height 3. Pallial sinus 'U' shaped and very deep. Habitat: Muddy.
2.	*Katelysia opima*	1. Shell is inflated, solid and thick with sooth surface. 2. Shell is slightly longer than height and margin rounded. 3. Pallial sinus is deep and apex of sinus bluntly angular. Habitat: Muddy.
3.	*Meretrix meretrix*	1. Shell solid, thick and triangular oval. 2. Striates are absent, surface gloomy and polished. 3. Pallial sinus very shallow. Habitat: Sandy.
4.	*Meretrix casta*	1. Shell thick, smooth, devoid of and sculeture and triangularly ovate. 2. Outer surface pale yellowish brown tinted with dark grey posteriorly and very fainted rayed with grayish radial lines.
5.	*Anadara granosa*	1. Shell thick, inflated and dark brown. 2. Taxodent dentition (small teeth in a single straight series) and about 20 elevated ribs bearing rectangular nodules.
6.	*Villorita cyprinoids*	1. Shell thick, ovately triangular with strong concentric ridges; ridges more strongly developed in the anterior half. 2. Pallial sinus small, lunule narrow and ligament large.

2. Clam population is indicative of the suitability of the site with particular to the tidal level, substratum and salinity.
3. Farm should be located in estuary, bays and other sheltered areas close to the shore.
4. About 1-2 hr exposure and low tide level is desirable as it is easy to remove the predators. If exposures are too long then it result in poor growth due to reduced feeding and in summer there may be mortality due to desiccations.

5. Clam farms must be located in the areas where there are a little wave actions and area prone to frequent changes of contour and vulnerable to pollution are avoided.
6. Salinity tolerance varies from species to species.

Table 12.3: Biology of Important Clam Species

Sl.No.	Species	Growth (mm)			Length at First Maturity (mm)	Spawning Season
		1st Yr	2nd Yr	Max Size		
1.	*M. meretrix*	29-47	61-63	91	21-26	March to June
2.	*K. opima*	22-26	40-43	N.A.	N.A.	Oct to Nov.
3.	*P. malbarica*	26-30	43-45	65	11-20	Oct to Feb.
4.	*M. casta*	42-43	N.A.	55	11-17	Throught year
5.	*A. granosa*	41-48	55-60	72	20-24	Throught year (can be 2 reproductive cycle/yr
6.	*V. cyprinoides*	30-34	39-42	N.A.	11 15	May to Aug/Sept and Jan to March

Seed Collection

In clam culture, seed collected from natural ground are used. In Maharashtra the clam's seed are collected from natural beds measuring in 10-25 mm in length, one month after spawning by using specially designed net, locally called as 'khonda' (Patil *et al.*, 2002).

Hatchery Production of Spat

The availability of seed from the wild is unpredictable, so CMFRI has been developed hatchery technology for clam seed production on commercial basis.

Mature clams of 35-45 mm shell length are used as brood stock. In clams, spawning occurs both at elevated water temperature of about 34°C and also at the lower temperature of about 24°C on transfer to the conditioning room, after the thermal shock. Spat settlement takes place between 7th and 26th day after spawning in different clam species studied. The clam spats attain 2-3 cm in length in the nursery. Survival rate of 15-20 per cent. In the hatchery the microalgae *Isochrysis galbana* is given as food to the larvae and mixed microalgae, reared in outdoor tanks as food to the spat (Narsimham and Laxmilatha, 1996).

Culture Method

For grow out culture, first ground should be leveled and cleaned of predators such as gastropods, crabs, starfishes etc. The movement of clams is limited or sluggish and in many areas fencing is not necessary. Synthetic fibre net pens are erected to protect clams against strong water current or bamboo stakes with nylon netting are used (Narsimham and Laxmilatha, 1996).

Harvesting

Clams are harvested after 5 to 6 months either by hand picking by use of hand operated dredge.

Packing

Clams, which are harvested, packed in 5 kg gunny bags. They can survive upto 12-15 hrs. The main market places for the clams are Mumbai, Pune, Thane and Kolhapur.

Prospectus in Clam Culture

1. Clams culture does not involve high labour or cost input.
2. They provided nutritious food for human consumption.
3. Clams feed on phytoplankton and detritus give high production per unit area.
4. Clam meat used as supplementary feed in semi-intensive shrimp farming.
5. Shells which rich in calcium carbonate also form the raw material in the manufacture of cement and lime based industries.

Exploitation

Clams are fished by men, women and children all along the west coast of India. They are collected from the intertidal region of about 4 meter depth. They are hand picked and also a hand operated dredge is used. Plank-built non powered boats are used for transport. Clams under one year age contribute in considerable numbers to the fishery. At many centers, November to April is the peak fishing season as the new recruits become available to the fishery.

References

Appukuttan, K.K., Aravindan, C.M., Yohanan, T.M. and Balasubramanium, N.K. 1999. Population dynamics of an exploited stock of the clam *Paphia malbarica* of Astamudi estuary (South India). In: Proceeding of the 4th Indian Fisheries Forum held at Kochi during Nov. 1996: 31-34.

Kripa, V. and Joseph, M. 1993. Clam fishery of North Vembanad lake. Mar. Fish. Info. Serv. T and E series, no. 119: 12-16.

Kripa, V. and Gopakumar, G. 1996. Utilization of clam meat as feed in the shrimp farms of Kerala. *J. Seafood Export*, 7(5): 7-14.

Narsimham, K.A. and Laxmilatha, P. 1996. Technology of clam culture. *Bull. Cent. Mar. Fish. Res. Inst.*, 48: 78-84.

Patil, S.V., Patil, Y.T. and Tandale, A.T. 2002. Clam resources, culture potential along the west coast of Maharashtra. *J. Seaf. Export*, 13 (1): 11-19.

Microbiological study was performed after removing 1g of the injured tissue from infected and uninfected fishes. The tissue samples were homogenized with sterilized seawater. 1ml of the serially diluted homogenized tissue samples was used for the isolation of total heterotrophic bacteria in Zobell marine agar, *Vibrio* count in thiosulphate citrate bile salts sucrose agar (TCBS), *Salmonella* in *Salmonella Shigella* agar (SS agar), *Klebsiella pneumoniae* in xylose lysine deoxycholate agar (XLD agar) and *Pseudomonas aeruginosa* in certrimade agar. Total counts of bacteria were expressed in terms of colony forming unit/g (CFU/g). Bacterial strains isolated from the aforementioned media were identified with Bergey's manual of determinative bacteriology (Buchanan and Gibbons, 1974) based on their biochemical and physiological characteristics. Identified strains were stored in Zobell marine agar slants at 4°C.

Results and Discussion

The ventral region of the infected fishes showed characteristics of skin lesions with red colour and continuous bleeding. The bacterial counts observed were: 6.6×10^8CFU/g for the total heterotrophic bacterial count in the parasite infected *Arius jella* whereas in the uninfected fishes it was only 1.10×10^2 CFU/g; 5.0×10^3CFU/g and 3.4×10^1CFU/ml for the total *Vibrio* count in infected and uninfected fishes respectively with the contribution of *V. cholerae*, *V. parahaemolyticus* and *V. vulnificus* estimated at 16.6 per cent, 30.3 per cent and 28.7 per cent respectively, in both cases; 6.0×10^2CFU/g for *S. typhimurium*, 6.0×10^1CFU/g for *K. pneumoniae* and 6.0×10^1CFU/g for *P. aeruginosa* found only in infected fish and absent in uninfected fish. Among the total bacterial isolates, *Vibrio* spp. were predominating with a percentage of 75.6 per cent, followed by *S. typhimurium* (9.0 per cent), *P. aeruginosa* (9.0 per cent) and *K. pneumoniae* (6.0 per cent) (Table13.1). Statistical analysis (t-test) of the results obtained from the resent study showed a high significance of $p<0.05$.

Table 13.1: Bacterial counts in infected *Arius jella*

THB	*(6.6X10⁸CFU/g)*	
Species	*No. of Strains*	*% Composition*
Vibrio cholerae	11	16.6
V. parahaemolyticus	20	30.3
V. vulnificus	19	28.7
Salmonella typhimurium	6	9.0
Klebsiella pneumoniae	4	6.0
Pseudomonas aeruginosa	6	9.0

Parasitic crustaceans are the largest fish parasites inducing significant damages to their hosts. They were involved in the spread of *Lymphocystis* disease (Nigrelli, 1950), *Icthyophorrus* contamination (Mann, 1970) and other secondary microbial infections. Overstreet and Howse (1977) opined that haemorrhagic lesions in the spotted gore parasitized by the cymothoid *Anilocra acuta* were subject to secondary infection. The cymothoid *N. phaiopleura* parasitizing *Arius jella* paved the way for access of pathogenic microbes.

Species of the genus *Nerocila* were reported infecting various families of marine fishes (Trilles, 1994), the parasitic damages including growth retardation, emaciation and frequently death of the fishes affected. Mucus overproduction has been noted in all parasitic diseases (Trilles, 1968, 1969; Romestand, 1978, 1979; Bragoni *et al.*, 1983, 1984; Rajkumar *et al.*, 2005 and 2007). Lesions observed on

Arius jella are similar to those reported by Williams and Williams (1978) on the sea catfish *A. felis* infested by *N. acuminata* or caused by other skin-dwelling crustaceans inducing localized epidermal destruction and hyperplasia.

Bacterial load in the infected tissue revealed that THB and *Vibrio* were higher. This may be attributed to the loss of intact skin in infected fishes, which allows interactions of the environmental microbes with the underlying tissue, with high lipid content. The higher THB and *Vibrio* counts in the infected region could be due to the severity of the lesions at that level as reported in the Creole fish parasitized by *N. acuminata* (Rand, 1986). In uninfected fishes, THB and *Vibrio* alone were observed in very low density. This may be due to the healthiness of the fish, which disturb the contact between microbes and fish, in addition of the continuous secretion of mucus in all regions.

However, the bacteria observed at the lesions level may also contribute to damage host tissues. Whereas it is not possible in uninfected fishes. Furthermore, localized loss of osmoregulatory skin function may occur. Finally, isopods: (1) alter skin pigmentation on current and old settlement and feeding sites, and (2) may mechanically impede swimming performance, so making infested fish more susceptible to predation and indirectly inducing an increased mortality in wild fish populations.

Hemorrhages on the body surface of gold fish infected by *V. cholerae* were reported by Reddacliff *et al.* (1993) and *V. vulnificus* induce a serious outbreak of disease on the cultured eels in Japan (Muroga *et al.*, 1976; Nishibuchi and Muroga, 1977, 1980; Nishibuchi *et al.*, 1979, 1980). *V. parahaemolyticus* is common in many coastal areas, freshwater bodies, plankton and fishes (Sarkar *et al.*, 1985; Ramamurthy and Nair, 2004).

The attempt of the present study is to achieve an approach of the extent of such secondary bacterial infections in the infected area. These bacteria can be opportunistic pathogens or produce secondary infections. The two dominant isolated *Vibrio* species are *V. angullarium* and *V. parahaemolyticus*. *V. angullarium* has been earlier reported as a potential fish pathogen by several authors (Kusuda, 1966; Hacking and Budd, 1971; Parker and Smith, 1984; Singh *et al.*, 1996). Vibriosis induced by *V. angullarium* affect a wide variety of fish species from brackish water to seawater (Loganathan, 1985). Harrell *et al.* (1976) reported that isolates of *V. anguillarium* are the most important aetiological agent of vibriosis. Infected areas are of greater significance because they may lead to the death of organism. Lightner (1977) specified that *V. parahaemolyticus* be among organisms inducing vibriosis in prawns and also associated with secondary infections. *V. parahaemolyticus* was also reported from cultured marine fishes (Sindermann, 1970; Raghukumar, 1971; Ravichandran *et al.*, 2001). However, *V. parahaemolyticus* has not been specifically identified as a pathogen of marine fish although Kusuda (1966) reported *Vibrios* inducing ulcerous diseases in cultured marine fish with close similarity to intestinal inflammation in man caused by *V. parahaemolyticus*. According to the above results, *V. parahaemolyticus* can also be considered as a significant pathogen in secondary infection.

Salmonella was reported in the gut of tilapia and carps (Iyer and Shrivastava, 1989; Ogbondeminu, 1993) and as a fish pathogen in salmon (Austin *et al.*, 1982). Occurrence of *Pseudomonas* in infected fish has been specified by Eddy and Jones (2002), Simidu and Tsukamoto (1985) and *Salmonella* by Karunasagar *et al.* (2004). *Pseudomonas* are everywhere in the aquatic environment, associated with both healthy and diseased fish (Evelyn and McDermott, 1961; Bullock and Sniesko, 1969). *P. anguilliseptica* and *P. fluorescens* were reported as a causative agent of red spot and generalized septicaemia in rainbow trout and most of the marine species (Austin and Austin, 1999). Presence of *K. pneumoniae* in a badly damaged fins and tail was recorded by Austin and Austin (1999).

The present study on parasite induced bacterial infection in wild *Arius jella* shows that such fish diseases are not specific to cultured organisms but also common in wild fishes.

Acknowledgements

We thank the authorities of Annamalai University for the facilities provided and to the Council of Scientific and Industrial Research (Govt. of India) for the financial support.

References

Austin, B. and D.A. Austin, 1999. Bacterial fish pathogens: Disease in farmed and wild fish. Springer Published in association with Praxis Publishing, Chichester, U.K. pp. 454.

Austin, B., M. Green and C.J. Rodgers, 1982. Morphological diversity among strains of *Yersinia ruckeri*. *Aquaculture*, 27: 73-78.

Bragoni, G., B. Romestand and J.P. Trilles, 1983. Parasitoses à Cymothoadien chez le loup (*Dicentrarchus labrax* Linnaeus, 1758) en élevage. II. Ecophysiologie parasitaire dans le cas de l'étang de Diana (Haute-Corse). *Ann. Parasitol. Hum. Comp.*, 58(6): 593-609.

Bragoni, G., B. Romestand and J.P. Trilles, 1984. Parasitoses à Cymothoadien chez le loup *Dicentrarchus labrax* (Linnaeus, 1758) en élevage. I. Ecologie parasitaire dans le cas de l'étang de Diana (Haute-Corse) (Isopoda, Cymothoidae). *Crustaceana*, 47(1): 44-51.

Buchanan, R.E. and N.E. Gibbons, 1974. Bergey's manual of determinative bacteriology, 8th Edition. The Williams and Wilkins Company, Baltimore. pp. 1246.

Bullock, G.L. and S.F. Sniesko, 1969. Bacteria in blood and kidney of apparently healthy hatchery trout. *T. Am. Fish. Soc.*, 98: 268-269.

Bunkley-Williams, L. and E.H.Jr. Williams, 1998. Isopods associated with fishes: a synopsis and corrections. *J. Parasitol.*, 84(5): 893-896

Eddy, S.D. and S.H. Jones, 2002. Microbiology of summer flounder *Paralichthys dentatus* fingerling production at a marine fish hatchery. *Aquaculture*, 211: 9-28.

Evelyn, T.P.T. and L.A. McDermott, 1961. Bacteriological studies of freshwater fish. I. Isolation of aerobic bacteria from several species of Ontario fish. *Can. J. Microbiol.*, 7: 375-382.

Grisez, L., M. Chair, P. Sorgeloos and F. Ollevier, 1996. Mode of infection and spread of *Vibrio anguillarum* in turbot *Scophthalmus maximus* larvae after oral challenge through live feed. *Dis. Aquat. Org.*, 26: 181-187.

Hacking, M.A. and J. Budd, 1971. *Vibrio* infection in tropical fish in a freshwater aquarium. *J. Wildlife Dis.*, 273-280.

Harrell, L., A.J. Novotny, M.H. Schiewe and H. Hodgins, 1976. Isolation and description of two vibrios pathogenic to pacific salmon in Puget Sound, Washington. *Fisheries Bull.*, 74: 447-449.

Heckmann, R. 2003. Other ectoparasites infecting fish; copepods, branchiurans, isopods, mites and bivalves. *Aquaculture Magazine*, 29 (6): 20-31.

Horne, M.T., R.H Richards, R.J Roberts and P.C. Smith, 1977. Peracute vibriosis in juvenile turbot *Scophthalmus maximus*. *J. Fish Biol.*, 11: 355-361.

Iyer, T.S.G. and K.P. Shrivastava, 1989. On the pattern of *Salmonella* serotypes in fishery products, frog legs and processing environments. *Fisheries Technol.*, 26: 131-136.

Kabata, Z. 1970. Crustacea as enemies of fishes. In: Diseases of Fishes I. (S.F. Snieszko and H.R. Axebroad Eds.), pp.1-171. T.F.H. Publications Inc. Jersey City. N.J.

Kaneko, T. and R.R. Colwell, 1973. Ecology of *Vibrio parahaemolyticus* in Chesapeake Bay. *J. Bacteriol.*, 113: 24-32.

Karunasagar, I., I. Karunasagar and A. Parvathi, 2004. Microbial safety of fishery products. Marine microbiology, facets and Opportunities. Nagappa Ramaiah, National Institute of Oceanography, Dona Paula, Goa, India. pp. 183-201.

Kusuda, R. 1966. Studies on the ulcer disease of marine fishes. The conference on marine microbiology. The first U.S Japan joint conference: pp. 1-3.

Lightner, D.V. 1977. Shrimp disease. In: Disease, diagnosis and control in North American marine aquaculture. (C.J. Sindermann, Ed.), Developments in Aquaculture and Fisheries Science, 6: pp.10-77. Elsevier Sci. Publ. Co., Amsterdam.

Loganathan, B. 1985. Bacterial and fungal diseases of fishes of Porto-Novo. Ph.D. Thesis, Annamalai University, Parangipettai, India. pp. 204.

Mann, H. 1970. Copepoda and Isopoda as parasite of marine fishes. Symposium on Diseases of Fish and Shellfish. Am. Fish. Soc., Special Publication, Washington, D.C., 5: 117-189.

Maxwell, J.G.H. 1982. Infestation of jack mackerel, *Trachurus declivis* (Jengus), with cymothoid isopod, *Ceratothoa imbricatus* (Fabricius) in Southeastern Australian waters. *J. Fish. Biol.*, 29(3): 341-349.

Muroga, K., H. Yasunobu, N. Okada and K. Masumura, 1990. Bacterial enteritis of cultured flounder *Paralichthys olivaceus* larvae. *Dis. Aquat. Org.*, 9: 121-125.

Muroga, K., M. Nishibuchiand and Y. Jo. 1976. Pathogenic *Vibrio* isolated from cultured eels. II. Physiological characteristics and pathogenicity. *Fish Pathol.*, 12: 147-151.

Nair, S., K. Kita-Tsukamoto and U. Simidu, 1988. Bacterial flora of healthy and abnormal chaetognaths. *Nippon Suisan Gakk.*, 54(3): 491-496.

Nigrelli, R.E. 1950. Lymphocystis disease and Ergasilid parasites in fishes. *J. Parasitol.*, 36: 36.

Nishibuchi, M. and K. Muroga, 1977. Pathogenic *Vibrio* isolated from cultured eels. III. NaCl tolerance and flagellation. *Fish Pathol.*, 12: 87-92.

Nishibuchi, M. and K. Muroga, 1980. Pathogenic *Vibrio* isolated from cultured eels. V. Serological studies. *Fish Pathol.*, 14: 117-127.

Nishibuchi, M., K. Muroga and Y. Jo, 1980. Pathogenic *Vibrio* isolated from cultured eels. VI. Diagnostic tests for the disease due to the present bacterium. *Fish Pathol.*, 14: 125-131.

Nishibuchi, M., K. Muroga, R.J. Seidler and J.L. Fryer, 1979. Pathogenic *Vibrio* isolated from cultured eels. IV. Deoxyribonucleic acid studies. *Bull. Jap. Soc. Scient. Fish.*, 45: 1469-1473.

Ogbondeminu, F.S. 1993. The occurrence and distribution of enteric bacteria in fish and water of tropical ponds in Nigeria. *J. Aquacult. Trop.*, 8: 61-66.

Overstreet, R.M. and R.D. Howse, 1977. Some parasites and diseases of estuarine fishes in polluted habitats of Mississippi. *Ann. N.Y. Acad. Sci.*, 298: 427-462.

Parker, M.T. and G. Smith, 1984. *Vibrio, Aeromonas, Plesiomonas, Compylobacter* and *Sprillum*. (Tropley and Wilson, Eds:), pp.1-562. Principles of bacteriology, Virology and immunity, (7th edition), 2.

Pederson, K., I. Dalsgaard and J.L. Larsen, 1997. *Vibrio damsela* associated with diseased fish in Denmark. *Appl. Environ. Microbiol.*, 63: 3711-3715.

Raghukumar, S. 1971. Studies on fungi in marine and brackish environments. Ph.D., Thesis, University of Madras. pp. 220.

Rajkumar, M. K.P. Kumaraguru vasagam, P. Perumal and J.P. Trilles, 2005. First record of *Cymothoa indica*, a cymothoid isopod parasite, from cultured catfish, *Mystus gulio* in India. *Dis. Aqua. Org.*, 65(3): 269-272.

Rajkumar, M., R. Thavasi, P. Perumal and J.P. Trilles, 2007. Parasite induced vibriosis in *Stolephorus commersonii*. *Res. J. Microbiol.*, 2(12): 972-977.

Ramamurthy, T. and G. Balakrish Nair, 2004. *Vibrio parahaemolyticus*: The threat of another Vibrio acquiring pandemic potential. Marine microbiology, facets and Opportunities. Nagappa Ramaiah, National Institute of Oceanography, Dona Paula, Goa, India. pp. 141-156.

Rand, T.G. 1986. The histopathology of infestation of *Paranthias furcifer* (L.) (Osteichthyes: Serranidae) by *Nerocila acuminata* (Schioedte and Meinert) (Crustacea: Isopoda: Cymothoidae). *J. Fish Dis.*, 9: 143-146.

Ravichandran S., A.J.A. Ranjit Singh and N. Veerappan, 2001. Parasite-induced *Vibriosis* in *Chirocentrus dorab* off Parangipettai coastal waters. *Curr. Sci.*, 80: 622-623.

Reddacliff, G.L., M. Hornitsky, J. Carson, R. Peterson and R. Zelski, 1993. Mortalities of gold fish, *Carassius auratus* (L.), associated with *Vibrio cholerae* (non-O1). *J. Fish Dis.*, 16: 517-520.

Romestand, B. 1978. Etude écophysiologique des parasitoses à Cymothoidae. Ph.D. Thesis, University of Montpellier II, France. 284 pp. I. XXXI.

Romestand, B. 1979. Etude écophysiologique des parasitoses à Cymothoadiens. *Ann. Parasitol. Hum. Comp.*, 54(4): 423-448.

Sarkar, B.L., G.B. Nair, A. K. Banerjee and S.C. Pal., 1985. Seasonal distribution of *Vibrio parahaemolyticus* in fresh water environs and in association with freshwater fishes in Calcutta. *App. Environ. Microbiol.*, 49: 132-136.

Sedano, J., I. Zorrilla, M.C. Morinigo, A. Vidaurreta, M.A. Bordas and J. Borrego, 1996. Microbial origin of the abdominal swelling affected farmed larvae of gilt-head sea bream, *Sparus aurata* L. *Aquacult. Res.*, 27: 323-333.

Simidu V. and K. Tsukamoto, 1985. Habitat segregation and biochemical activities of marine members of the family Vibrionaceae. *Appl. Environ. Microbiol.*, 50: 781-790.

Simidu, U. and K. Aiso, 1962. Occurrence and distribution of heterotrophic bacteria in sea water from Kamogawa Bay. *Bull. Jap. Soc. sci. Fish.*, 28: 1133-1141.

Simidu, U., E. Kaneko and N. Taga, 1977. Microbiological studies of Tokyo Bay. *Microbial Ecol.*, 3: 173-191.

Sindermann, C.J. 1970. Principal diseases of marine fish and shellfish. Academic press, New York and London. pp. 155.

Singh, B.R., M.P. Saxena, S.B. Kulshreshtha and K.N. Kapoor, 1996. A note on prevalence of *Vibrios* in marine fish and shellfishes. *Indian J. Fish.*, 43: 407-410.

Trilles, J.P. 1968. Recherches sur les isopodes Cymothoidae des côtes francaises. Ph.D. Thesis, University of Montpellier II, France. 793 pp.

Trilles, J.P. 1969. Recherches sur les isopodes 'Cymothoidae' des côtes francaises. Apercu génèral et comparatif sur la bionomie et la sexualité de ces Crustacés. *Bull. Soc. Zool. Fr.*, 94(3): 433-445

Trilles, J.P. 1979. Les Cymothoidae (Isopoda, Flabellifera, parasites de poissons) du Rijksmuseum van Natuurlijke Historie de Leiden II. Afrique, Amérique et régions Indo-ouest-pacifiques. Zoöl Meded., Leiden., 54: 245-275.

Trilles, J.P. 1994. Les Cymothoidae (Crustacea, Isopoda) du Monde (Prodrome pour une faune). *Stud. Mar.*, 21/22 (1-2), 1991: 5-288.

Williams, E.H. Jr. and L.B. Williams, 1978. Cymothoid isopods of some marine fishes from the northern Gulf of Mexico. *Northeast Gulf Sciences*, 2: 122-124.

Zobell, C.E. 1946. Marine microbiology. A monograph on hydrobacteriology. Chronica Botanica. Waltham, Massachusetts Estados Unidos de Norteamerica.

Chapter 14

Assessment of Polyculture Trial of Giant Freshwater Prawn (*M. rosenbergii*) with *O. mossambicus* (Tilapia) and Carps in the Ponds of Aquafarmer

P.E. Shingare, G.A. Shirgur and V.B. Mehta

Introduction

Extensive pond culture trials were undertaken by Taraporevala Marine Biological Research Station, Bombay for the last one decade in low saline ponds of Agriculture Research Station, Panvel, freshwater ponds at Tata Electric Company farm, Lonavala and the various ponds belonging to private Aquafarmers. The polyclture trials comprising carps, Tilapia and Prawn was undertaken to assess. Productivity in relation to qualitative and quantitative input of seed and organic substances. The biomass productivity was in the range of 1000-1300 kg/ha. All these details are published in the review papers (Shirgur, 1993 b; Shirgur, 1994 a). In view of availability of natural seed during September-October period from several sites of westward flowing rivers of Konkan region of Maharashtra many Aquafarmers have started undertaking monoculture and polyculture trials. The present studies were therefore undertaken in two pons of Aquafarmer at Niphan village near Khopoli, district-Raigad for evaluation of extensive culture methods.

Materials and Methods

All the relevant details are shown in Table 14.1. The information shown is self- explanatory.

Table 14.1: Norms for Conducting Polyculture Trials of *M. rosenbergii* (Giant Freshwater Prawn) with *Oreochromis mossambicus* (Tilapia) and Carps from Two Ponds of the Aquafarmer at Niphan Village Near Khopoli, Distract–Raigad (Maharashtra) and Results

Norms

Sl.No.	*Particulars*	*Concerned Details*					
		Pond No.1			*Pond No.2*		
1	*2*	*3*			*4*		
1.	Location of the ponds	Niphan near Khopoli dist: Raigad			Niphan near Khopoli dist: Raigad		
2.	Waterspread area and mean depth (m)	0.20 (2.25)			0.15 (1.75)		
3.	Duration of culture	9 ½ months			9 ½ months		
4.	Specieswise weed stocked	Nos.	N per cent	Av. wt (g)	Nos.	N per cent	Av. wt (g)
	(a) Catla	5000	8	2.750	2667	18.18	2.750
	(b) Rohu	5000	8	2.190	667	4.55	2.190
	(c) Mrigal	–	–	–	667	4.55	1.120
	(d) Cyprinus	2500	4	2.500	1333	9.09	–
	(e) Tilapia	–	–	–	1333	9.09	2.500
	(f) Prawn	50000	80	9.280	9000	54.54	0.200
	Total	62500	100	–	14667	10000	–
5.	Organic substances used besides SSP applied (Kg) and per cent composition	Qty		Per cent Comp.	Qty		Per cent Comp.
	(a) Wheat floor	2560		48.49	2133.333		47.62
	(b) Offals (dry wt.*)	960		18.18	853.200		19.04
	(c) Wet RCD (dry wt.*)	800		15.15	853.400		19.05
	(d) Hotel Waste (dry wt.*)	960		18.18	646.00		14.29
	Total	5280		100.00	4480.00		100.00
	(e) S.S.P.	160		–	213.00		–
	Total	814.111		100	2967		100
6.	Mode of distribution in relation to CBW of fish biomass every alternate day	All as direct feeds			All as direct feeds		

Contd...

Table 14.1–Contd...

Results

Sl.No.	*Particulars*	*Concerned Details*					
		Pond No.1			*Pond No.2*		
7.	Specieswise harvesting details	Nos.	N per cent	Av. wt (g)	Nos.	N per cent	Av. wt (g)
	(a) Catla	1750	15.30	1.000	1213	18.53	0.700
	(b) Rohu	2000	17.48	0.300	467	7.13	0.210
	(c) Mrigal	–	–	–	420	6.42	0.195
	(d) Cyprinus	1100	9.62	0.500	167	2.55	0.490
	(e) Tilapia	–	–	–	2267	40.74	0.125
	(f) Prawn	6590	57.60	0.055	1613	24.63	0.050
	Total	11440	100.00	–	6547	100.00	–
8.	Percentage Survival, Total wt. (kg) and respective per cent Biomass	Per cent Survival	Total Wt. (Culture Period) kg.	Respective Per cent Biomass	Per cent Survival	Total Wt. (Culture Period) kg.	Respective Per cent Biomass
	a. Catla	35	1750.00	53.64	45.48	850.100	55.71
	b. Rohu	40	600.00	18.39	70.00	98.070	6.43
	c. Mrigal	–	–	–	62.96	81.900	5.37
	d. Cyprinus	44	550.00	16.86	12.53	81.830	5.36
	e. Tilapia	–	–	–	–	333.375	21.85
	f. Prawn	13.10	362.500	11.11	20.16	80.650	5.28
	Total	–	3262.500	100.00	–	1525.925	100.00
9.	Gain in wt. (a), Biomass; feed (b)		a 3221.55	b 1:1.61		a 1510.801	b 1:2.93

Values expressed as dry weight equal to 20 per cent of wet weight.

Figures are shown per hectare basis.

Abbreviations: Nos.: Numbers; N per cent: Numerical percentage; Av.wt.: Average weight; Qty: Quantity; Per cent comp; Percentage composition; SSP: Single Super Phosphate; wt.: Weight; RCD: Raw Cattle Dung; CBW: Collective Body Weight.

Results, Discussion and Conclusions

The results are presented under the Table 14.1. Considering these results in relation to the previous work following assessment are made.

1. Under polyculture trials prescribed stocking rate of prawn seed has been around 10,000/ha. However the Aquafarmer has stocked very heavily in pond No.1. As a result percentage survival in pond No.1 has been very low being around 13.10, with better biomass percentage around 11.11 as compared with half value in pond No.2.

2. In pond No.1 rate of total organic input was higher being around 5280 kg as compared with 4480 kg in pond No.2 Moreover in pond No.1, aeration was provided.
3. Among the organic feeds, wheat flour, wet raw cattle dung etc were acting as phased fertilizers, thereby enhancing zooplankton production especially favouring Indian major carps. The other feeds particularly the offals were suitable for faster rate of growth; for prawns. The other direct fees with high nutritive value were conducive for overall productivity in both the ponds. The results in pond No.1 were far better than in case of pond No.2 with respective Feed Conversion Ratio (F.C.R.) values being around 1:1:61 and 1:2.93 at the cost of very high seed stocking density in pond No.1.
4. It is to conclude that polyculture with Indian major carps, Cyprinus and Tilapia is feasible enterprise. However, optimal prawn seed stocking rate around 10,000 and others around 5000 per hectare needs to be followed for better productivity, percentage survival and overall economy.

Acknowledgements

The authors are indebed to Hon. Vice- Chancellor and Director of Research, Dr.B.S.Konkan Krishi Vidyapeeth, Dapoli, Dist: Ratnagiri for encouragement. The authors are thankful to the Associate Dean, College of Fisheries, Ratnagiri for overall help to conduct experimental work. The authors are also great ful to Shri Vasant Tilak, Aquafarmer for providing pond facilities.

References

Shirgur, G.A.1993 b. Mass culture techniques of feed micro-organisms during post- larval and nursery phases of grant freshwater prawn, *M.rosenbergii* (M.S.), Pb.*fish Bull*, XVII (2): 1-13.

Shirgur, G.A.1994 a.Some aspects of seed procurement, rearing and pond culture of *M. rosenbergii* Proceedings of the workshop on status of freshwater prawn farming in India, 17-18 Mar. 1994: 93-102.

Chapter 15

Role of Women in Aquaculture

K. Sree Latha

Women in aquaculture make a substantial portion of Indian rural population like men, they are engaged in farm operations like seed collection, maintenance of ponds, manuring, Intercultivation, harvesting and harvest technology, processing and preservation and marketing. Women participated in large numbers and played a significant role in aquaculture produce. Their participation is increasing in spite of scientific and technological developments. Recent studies highlight that women in India are major produces of food in terms of value, volume and number of hours worked. Nearly 63 per cent of all economically active men are engaged as compared to 78 per cent of women. In India women forces are overwhelmingly engaged in aquaculture as both main and marginal workers from ancient times. There is no denying fact that, the rural women have been contributing largely to the country's economy, which is mainly agriculture based and next to it is aquaculture. Our country stands second in Asia continent and contributes 9 per cent of the worlds total aquaculture. Their share is of labour is particularly significant in the context that they actually perform their work in addition to long and arduous work in house hold maintenance, parental care, cooking, fuel and wood collection, fodder collection, fetching drinking water etc. However, unfortunately this the sector where the women play a role as unpaid labour in productive activities and is most prominent. At present, women mostly participate in pond construction activities, selection of species for culture-, seed, fry, fingerling collection, sorting and identification, their supply, maintenance of hatching pits, nursery ponds rearing ponds and stocking ponds, checking the water supply and restoration of culture ponds. Control of aquatic vegetation, stripping and induced breeding activities. Constant monitoring (including prevention of pond water from adverse effects of pollutants). Pond Fertilisation and manuring, removal of weeds, taking measures for protection of culture fish from predators (especially birds and mammals), netting activities, transportation of catch to local and distant markets for sale. After the harvest the produce is processed kept in cold storage units for export, women are also involved in making nets and hapas, collection of wood and boat building, rearing fish feed organisms(plankton), feed preparations, making nests for brooders(spawning), preparation of cages and pens for capatability of culture fish, preservation of fish and fish products by drying, smoking, salting, smoking, canning, where the women are involved muchin such activities. Besides these, the women are also involved in protection of culture ponds from natural disasters, preservation of germ plasm, plantations along the culture pond in integrated aquaculture, other conservation activities for the protection of fauna of the region. They have played an important role in protection of aquatic environments. As there is an increasing demand for fish and fish products, fast food outlets (Non-vegetarian) established by women and fisher women cooperatives are gaining popularity in this cosmopolitan world. Not only as head loaders, the women are doing institutional marketing too. Though they are not earning considerable amount of money, but were very conscious about the benefits of aquaculture produce. And take active part in raising, harvesting and marketing the fish and their products from natural aquatic habitats and establishing ponds in their villages.

The role of aquaculture is changed dramatically after the advent of science and technology, will provide more than 50 per cent by 2025. In farming practices sustaining the tempo of aqua cultural produce in the country, farm women will have to be given more prominence in aqua cultural and allied vocation through provision of resources, education and training in modern and technologically sound economically viable aquaculture gadgets should be popularized in rural areas. The application of science and technology will improve the status of women in the society. Her role should not be confined to household work but also she should as well be made an important citizen of the society. Of course in India women cannot be ignored as the aquaculture is an integrated part of agriculture, the important segment of our rural population and hence in nation building.

Chapter 16

Polyculture Trials of *Lates calcarifer* (Jitada) with *O. mossambicus* (Tilapia) and Carps in the Station Pond and in Private Aquafarmer

P.E. Shingare, G.A. Shirgur and V.B. Mehta

Introduction

Traditional polyculture of Jitada with Tilapia, Indian major carps and common carp is being practiced in the ponds of Raigad and Thane districts as per the survey undertaken by Taraporevals Marine Biological Research Station, Bombay (Belsare *et al.*, 1987; Shirgur, 1990 b).

Accordingly maximum productivity around 3000 kg/ha/year was reported. Therefore, with a view to evaluate such traditional practices, the present work was undertaken in two ponds, located at Agricultural Research Station, Panvel and nearby Dhakati Jui pond in Raigad district of Maharashtra State. Earlier such trials (Shirgur, *et al.*, 1993) had indicated about better response of Jitada and Indian major carps for organic inputs under phased fertilization technique. The other culture trials using Jitada and Tilapia also has revealed about suitability of both phased fertilization technique and direct feeding methods besides combined method (Shirgur, 1990 c, Singh and Shirgur, 1994).

Materials and Methods

The methodological norms are shown in Table 16.1. This information is self-explanatory.

Table 16.1: Norms for Conducting Polyculture Trials of *Lates calcarifer* (Jitada) with *Oreochromis mossambicus* (Tilapia), Indian Major Carps and Common Carp in the Ponds and Results Obtained

(A) Norms

Sl.No.	*Particulars*	*Concerned Details*					
		ARS–Pond No.1			*Dhakati Jui Pond*		
1	*2*	*3*			*4*		
1.	Location of the ponds	Panvel Dist: Raigad			Dhakati Jui (Uran) Dist: Raigad		
2.	Waterspread area and mean depth (m)	0.737 (1.40)			0.12 (1.75)		
3.	Duration of culture	9 months			10 months		
4.	Specieswise weed stocked	Nos.	N per cent	Av. wt (g)	Nos.	N per cent	Av. wt (g)
	(a) Jitada	1493	19.72	5.400	1042	20	6.00
	(b) Catla	800	10.57	2.750	208	4	100.00
	(c) Rohu	800	10.57	2.190	208	4	100.00
	(d) Mrigal	407	5.38	1.112	–	–	–
	(e) Cyprinus	–	–	–	833	16	3.000
	(f) Tilapia	4070	53.76	0.489	2917	56	0.489
	Total	7570	100.00	–	5208	100	–
5.	Organic substances used besides SSP applied (Kg) and per cent composition	Qty		Per cent Comp.	Qty		Per cent Comp.
	(a) DOCG	325.644		40	–		–
	(b) RB	325.644		40	1666.666		27
	(c) Wet RCD (dry wt.*)	162.822		20	500.000		32.50
	(d) Small frogs (dry wt.*)	–		–	400.00		32.50
	(e) Small local fishes (dry wt.*)	–		–	400.00		8
	(f) S.S.P	19.946		–	166.660		–
	Total	814.111		100	2967		100
6.	Mode of distribution shown in relation to CBW of fish biomass every alternate day.	50 per cent DOCG, and 100 per cent in liquid (phased) Remaining as direct feed		100 per cent RCD SSP applied form fertilization)	All as directed feeds		

Contd...

Table 16.1–Contd...

Results

Sl.No.	*Particulars*	*Concerned Details*					
		ARS–Pond No.1			*Dhakati Jui Pond*		
7.	Specieswise harvesting details	Nos.	N per cent	Av. wt (g)	Nos.	N per cent	Av. wt (g)
	(a) Jitada	421	36.96	0.303	958	28.32	0.939
	(b) Catla	81	7.11	1.125	208	6.15	1.200
	(c) Rohu	176	15.45	0.410	133	3.93	0.750
	(d) Mrigal	54	4.75	0.500	–	–	–
	(e) Cyprinus	–	–	–	417	12.33	0.800
	(f) Tilapia	407	35.73	0.098	1667	49.27	0.150
	Total	1139	100.00	–	3383	100.00	–
8.	Percentage Survival, Total wt. (kg) and respective per cent Biomass	Per cent Survival	Total Wt. (Culture Period) kg.	Respective Per cent Biomass	Per cent Survival	Total Wt. (Culture Period) kg.	Respective Per cent Biomass
	(a) Jitada	28.19	127.540	34.56	91.93	900.000	49.09
	(b) Catla	10.12	91.600	24.82	100.00	250.00	13.64
	(c) Rohu	22.00	82.770	22.43	63.94	100.00	5.45
	(d) Mrigal	13.17	27.140	7.35	–	–	–
	(e) Cyprinus	–	–	–	50.06	333.300	18.18
	(f) Tilapia	10.00	40.000	10.84	57.14	250.00	13.64
	Total	–	369.050	100	–	1833.300	100
9.	Total Wt. rate for 10 months (a) Gain in wt. (b), Biomass, feed (c)	a 394.00	b 354.6	C 1:2.29	a 1781.45	b 1781.45	c 1:1.66

* Values expressed as dry weight equal to 20 per cent of wet weight.

Figures are shown per hectare basis.

Abbreviations: ARS: Agril. Research Station; Nos: Numbers; N per cent: Numerical percentage; Av.Wt.: Average Weight; Qty: Quantity; Per cent Comp: Percentage Composition; DOCG: De-oiled Cake of Groundnut; RB: Rice bran; SSP: Single Super Phosphate; RCD: Raw Cattle Dung; Wt.: Weight; CBW: Collective Body Weight.

Results, Discussion and Conclusion

The salient results upon harvesting at the end of 9 to 10 months culture duration are shown in Table 16.1. All these values were therefore evaluated for the common duration of 10 months for both the ponds. Accordingly following conclusions were drawn.

1. The culture method adopted in Agril. Research Station Pond-1 has been sound as can be seen form previous trials. However lesser productivity during the present trial appears to be due to one third rate of total input of organic substances in relation to Dhakati Jui pond.

Futher, the seed size of Indian major carps was comparatively smaller in Agril. Research Station-1 pond. Pouching was also a major cause in this pond apparently resulting in comparatively lesser producitivity.

2. In Dhakati Jui pond because of direct feeds comprising live feed organisms such as a small frogs, small local fishes etc. percentage survival of Jitada has been very high around 91.93 besides higher biomass percentage around 50. The watch ans ward by the Aquafarmer has also resulted in overall better yield.
3. It can thus be concluded that polyculture trials comprising Jitada, Tilapia and Carps is feasible under combined method of phased fertilization and direct organic feeding and mainly direct feeding comprising the various live organic feed organisms.

Acknowledgements

The authors are indebed to Hon. Vice- Chancellor and Director of Research, Dr.B.S.Konkan Krishi Vidyapeeth, Dapoli, Dist: Ratnagiri for encouragement. The authors are thankful to the Associate Dean, College of Fisheries, Ratnagiri and Senior Scientific Officer, MBRS, Ratnagiri for overall help to conduct experimental work. The authors are also great ful to Shri.Tukaram Gharat, Aquafarmer for providing pond facilities.

References

Belsare, S.G., Shirgur, G.A., and Indulkar, S.T. 1987.Traditional culture practices of *L. Carcarifer* (Bloch) in North Konkan region of Maharashtra. India. Abstr. *First Indian Fisheries Forum,* Manglore, Dec. 4-8-1987.

Shirgur, G.A.1990b. On lates Calcarifer culture trials to the ponds of Khar Land Region by phased fertilization technique Pb. *Fishing Chimes,* 10 (2): 15-20

Shirgur, G.A.1990c. *On lates Calcarifer* culture trials in the ponds of Khar Land Region by phased fertilization technique Pb. *Fish,* Bull. Vol. XV, No.2 July-Dec, 1991. 15-20.

Shirgur, G.A., Singh, R.K., and Siddiqui, S.Q.1993. On pond culture trials of *L. Calcarifer* a seabass with *O. Mossambicus* (Tilapia) under phased fertilization and direct feed technique at Khar Land Research Station, Panvel–Raigad (m.S.) Presented for the *Nat. Conf. on Recent Trends in Aquaculture, and Environment,* Bhopal. Sept. 1993.

Chapter 17

Seasonal Variations in the Haematological Parameters of Three Species of the Genus *Channa* from *Khandala* Reservoir, Dist. Osmanabad (MS)

K.R. Reddy, M.G. Babare and M.V. Mote

Introduction

The blood in fishes is red in colour and consists of fluid plasma and blood cells. In poikilothermic animals the blood plays an important role for maintaining the physiological status of the body in relation to changing environmental factors during different seasons of the year.

During monsoon drastic changes occur in physico-chemical parameters of the water. Thus the haematological parameters are greatly influenced by such changes in environmental parameters. In general during monsoon the fishes grows faster. This can be detected by knowing the optimum level of all haematological parameters. The changes in the body fluid physiology can be understood by haematological study. Such type of study has greater importance in the aquacultural practices. It provides valuable information in the health management of fishes. Earlier Siddiqui *et al.,* (1970), Srivastava and Agarwal (1981), Joshi (1989), Pai and Shanbagh (1989) were also carried out the work on fish haematology

Materials and Methods

The fishes were collected with regular intervals from *Khandala* reservoir and brought to the laboratory in live condition. A caudal cut was made to collect the blood from aorta. The blood is collected with help of syringe. Then the blood was transferred to a container, having EDTA as an anticoagulant. Clean slides were taken and blood smears were prepared then stained with Wright and Leishman's stain for morphological and differential studies.

Results and Discussion

After taking the precautions regarding the selection of fishes *viz.*, healthy mature fish, devoid any type of disease and body injury, the blood smears were prepared. The values were estimated in the three seasons *viz.*, winter, summer and monsoon. During the study period the values were recorded for different haematological parameters are as follows.

Channa punctatus

The polymorphic erythrocytes were observed. They were circular, oval and spindle shaped having the diameter ranged between 7.7 to 8.2 microns. The maximum diameter recorded in the monsoon as 8.2 microns. Total Erythrocyte Count (TEC) was less in monsoon and more during summer. In the monsoon it was $1.02 \times 10^6/mm^3$ and during summer, it was $1.26 \times 10^6/mm^3$. The haemoglobin content was maximum during summer as 8.1 gm per cent and minimum during monsoon 6.0 gm per cent.

Channa striatus

In this species also the erythrocytes were polymorphic nature. The diameter of the erythrocytes ranged between 6.2 to 6.8 microns. The maximum diameter was observed in monsoon as 6.8 microns and minimum in summer 6.2 microns. The Total Erythrocyte Count (TEC) was less in monsoon as $1.11 \times 10^6/mm^3$ and more in summer as $1.32 \times 10^6/mm^3$. The haemoglobin content was maximum during summer as 8.4 gm per cent and minimum during monsoon 6.2 gm per cent.

Channa gachua

The polymorphic nature of the erythrocytes are observed. The diameter of the erythrocytes ranged inbetween 5.8 microns to 6.4 microns. The maximum diameter was observed in monsoon as 6.4 microns and minimum in summer as 5.8 microns. The Total Erythrocyte Count (TEC) was less in monsoon $1.16 \times 10^6/mm^3$ and more in summer as $1.41 \times 10^6/mm^3$. The haemoglobin content was maximum during summer 8.6 gm per cent and minimum during monsoon 6.4 gm per cent.

During the different seasons of the year, the physico-chemical parameters of water shows fluctuations. These fluctuations shows drastic effect on the haematological parameters of the fish. Increase in the diameter of the erythrocyte was observed during monsoon in all the three species due to adequate quantity of dissolved oxygen in the surrounding environment. During monsoon due to addition of rain water, surrounding medium of a fish becomes hypotonic. In such environment the body fluid gets diluted and causes the volume of erythrocyte to increase and become turbid in codition. It is also observed by Siddiqui *et al.* (1970), Joshi and Tandon (1977), Srivastava and Agarwal (1981). In all the three species the increase in the cell size leads to decrease in the cell count and haemoglobin content of a fish.

Acknowledgement

Authors are very much thankful to Management, Principal and Head Zoology Departement of Arts, Science and Commerce College, Naldurg, Distt. Osmanabad for providing the necessary facilities during the work.

References

Joshi B.D. and R.S. Tandon.1977. Seasonal Variations in the haemotological values of fresh water fishes. *Comp. Physiol. Ecol.*, 2(1): 22-26,

Joshi B.D. 1989.Physico-Pathological Studies of the various blood parameters of different vertebrate group. *Pro.National symposium Animal Haedatop*. 17-23

Shrivastava A.K. and U.Agrwal.1981.easonal and dirunal variations of blood cell types in fresh water teleost, *colisa fasciatus.Comp.physical Ecol.* 6(1): 19-24.

Siddiqui A.Q., S.S. Khan and S.M. Naseem.1970.Differential blood cell counts in there species of major carps. *Current Sci.* 39 (4):86-87

Yeragi S.G. and R.S. Landhe.2004. Seasonal Variations in the haematological parameters of *Catla Catla* and and *Tilapia mossambica* of phirage Kharbav Lake Distt. Thane. *J. Comp. Toxical Physist Vol.* 1 (III and IV) 265-266.

Chapter 18

The Saline Soil Resources of Maharashtra and its Suitability for Aquaculture

S.S. Dhane, P.E. Shingare and S.J. Meshram

Area of Coastal and Inland Saline Soils in Maharashtra

Coastal

The Maharashtra State has 720 km. of coastal length with 54 creeks. It comprises the districts of Thane, Raigad, Ratnagiri and Sindhudurg. The coastal saline soil occur in these districts due to periodical inundation of cultivated land by creek / sea water during high tide. As per the master plan of development of coastal saline soils of Maharashtra, the total area estimated to be 65,465 ha (Anonymous, 1990). The break up is given in Table 18.1.

Table 18.1: Area Under Coastal Saline Soils in the Maharashtra State

District	*No. of Creeks*	*Area (ha)*
Thane	12	20,795
Raigad	07	31,800
Ratnagiri	18	5,770
Sindhudurg	17	7,100
Total	54	65,465

Inland

In some part of Maharashtra inland saline soils are developed due to continuous irrigation to the crop like sugarcane. In this connection 188.90 lakh ha. area is surveyed out of which 4.03 lakh ha are estimated to be in Maharashtra. The total Division wise breakup of inland saline soil in Major and Medium irrigation dam is given in Tables 18.2, 18.3 and 18.4.

Table 18.2: Divisionwise Break Up of Inland Saline Soils

Sl.No.	Division	Total Area (ha)
1.	Pune	17447
2.	North Maharashtra (Nashik)	3735
3.	Aurangabad	4651
4.	Nagpur	876
5.	Amravati	467
6.	Konkan (Coastal saline)	65445
	Total	27976

Source: D.I.R.D., Pune.

Table 18.3: Area of Inland Saline Soils of Private and Co-operative Irrigation dam

Sl.No.	Division	Total Area (ha)
1.	Sangali	7517
2.	Kolhapur	2399
3.	Satara	247
4.	Pune (daund)	783
	Total	10586

Source: D.I.R.D., Pune.

Table 18.4: Inland Saline Soils Area Around the Canals of Large Irrigation Project

Sl.No.	No. of Dam	Irrigation Area (ha)	Total Water Logged Area (ha)	Total Inland Saline Soils Area (ha)
1.	53	1284.071	15.605	12.580

Source: D.I.R.D., Pune.

Agro-climatic Zones

Coastal Saline Soils

The coastal saline soils are spread over in two agro–climatic zone of the state. The very high rainfall non-lateritic soil zone (VRN) comprises the whole of Thane district and northern part of Raigad district, where the soils are derived from the basaltic parent material. While the very high rainfall lateritic soil zone (VRL) covers the Ratnagiri and Sindhudurg districts. In coastal areas, the rains received from the south west monsoon are spread over from June to October with peak period between July and August. The total annual rainfall varies from 3000-4000 mm. The relative humidity ranges between 75-80 per cent. The month of May is hottest and has mean temperature of 35.5 $°_C$ (Sahu *et al.*, 1981). In spite of such heavy rainfall, occurrence of dry spells of 13, 17 and 31 days which occur in the month of August, September and October respectively. The above data shows that there is need for designing water storage reservoirs to harness rainwater for protective irrigation in such soils.

Inland Saline Soils

The area in Pune region comes under transitional zone with annual rainfall of 1456 mm. Marathwada region has assured rainfall with Kharif croppring, annual rainfall 857 mm. Vidarbha region comes under moderaterly high rainfall with trap soil, and rainfall is of 1088 mm. Dhule, Ahmednagar, Solapur and Sangli come under scarcity zone, aird to semiarid, having annual rainfall of 588 mm. A Jalgaon district area comes under zone of assured rainfall with Kharif and rabi cropping, having annual rainfall 730 mm. (Patil and Sonar, 1994).

Physico-chemical Properties

Coastal Saline Soils

Joshi (1985) studied physico–chemical properties of coastal saline soils. It was showed that soil in VRN zone had fine texture and poor hydraulic conductivity while that from VRL zone had coarse texture and moderate hydraulic conductivity. The characteristics are shown in Table 18.5.

Table 18.5: Physico-chemical Properties of Coastal Saline Soils of Maharashtra

Sl.No	*Soil Properties*	*VRN*	*VRL*
I	Physical properties		
	Sand (per cent)	18.10	51.43
	Silt (per cent)	38.23	22.91
A	Clay (per cent)	43.67	25.63
	Bulk density (Mg m^{-3})	1.26	1.29
	Hydraulic conductivity	0.25	1.26
	Total porosity (per cent)	48.48	50.60
	Available moisture (per cent)	17.93	13.62
II	Chemical properties		
	PHs	6.80	5.60
	Ece (ds m^{-1})	25.40	11.90
	SolubleCa^{2}++ Mg^{2}+ (meL^{-1})	145.3	59.40
	Soluble Na^{+} (MeL^{-1})	243.30	117.60
	Soluble K^{+} (meL^{-1})	2.90	2.0
	Soluble C1^{+}(meL^{-1})	319.80	127.2
	Soluble SO_4^{2-} (meL^{-1})	74.50	31.90
	Soluble HCO_3^{-} (meL^{-1})	12.60	15.70
	SAR	28.14	20.07

Contd...

Table 18.5–Contd...

Sl.No	Soil Properties	VRN	VRL
	CEC cmol (p^+) kg^{-1}	43.37	26.60
	Exchangeable Na^+ cmol (p^+) kg^{-1}	11.15	5.20
	ESP	25.70	25.14
	Organic carbon g kg^{-1}	4.8	7.7
	Available P_2O_5 kg ha^{-1}	75.10	53.97
	Available K_2O kg ha^{-1}	2533.60	1072.83
	$CaCO_3$ (per cent)	5.0	2.7
	DTPA extractable Zn mg kg^{-1}	1.73	1.75
	DTPA extractable Fe mg kg^{-1}	9.77	28.71
	DTPA extractable Cu mg kg^{-1}	6.80	8.60
	DTPA extractable Mn mg kg^{-1}	35.73	36.94
	DTPA extractable B mg kg^{-1}	1.39	0.98

Inland Saline Soils

The inland saline soils were categorized as saline, saline–sodic and sodic based upon pH, Ece and ESP values which are given in Table 18.6 (Patil and Sonar, 1994) and (More *et al.*, 1987).

Table 18.6: Categorisation of Inland Saline Soils

Nature of Salt Affected Soils	pH	Ece	ESP
Normal	<8.5	<4	<15
Saline sodic	>8.5	>4	>15
Sodic	>8.5	<4	>15
Saline	<8.5	>4	<15

The saline water resources in inland region is mainly due to natural origin of salt affected soils and availability of ground saline water and resultant water logging of lands due to continuous irrigation to crops like sugarcane. A salt affected soil is developed when salts are accumulated in the root zone to such an extent that they adversely affect yield of normal crop.

Saline Soils

These soils are dominated by chloride and sulphate salts of calcium, magnesium and sodium. pH is below 8.5. The electrical conductivity at 25°C is above 4 mmhos/cm. Where as Exchangeable Sodium Percentage (ESP) is less than 15.

Saline Sodic and Sodic Soils

These soils are dominated by carbonate and bicarbonate of sodium. The Exchangeable Sodium Percentage (ESP) is more than 15. pH is more than 8.5. While Electric Conductivity (EC) is more than 4 in saline sodic soil and less than 4 is sodic soil. The physico-chemical properties are shown in Table 18.7.

Table 18.7: Physico-chemical properties of Inland Coastal Saline Soils of Maharashtra

Sl.No	*Soil Properties*	*Average Values*
1.	Physical properties	
	Sand (per cent)	22.4
	Silt (per cent)	15.4
	Clay (per cent)	61.6
2.	Chemical properties	
	pH	8.2
	EC MS/CM	0.22
	Organic carbon (per cent)	0.54
	CaC03 (per cent)	6.3
	CEC me/100g	58.4
	Available Nitrogen (N)	205
	Available Phosphorus (P)	8.72
	Available Potassium (K)	533
	ESP	14.31
	Soluble C03- (MeL-1)	2.00
	Soluble HC03- (Mel-1)	10.13
	Soluble Cl- (MeL-1)	23.23
	Soluble S04- (MeL-1)	19.68
	Exchangeable Ca (Me/100g)	18.64
	Exchangeable Mg (Me/100g)	10.35
	Exchangeable Na (Me/100g)	05.77
	Exchangeable K (Me/100g)	5.44

Coastal Saline Soils Vs Inland Saline Soils

The Inland saline soils were clayey in texture. The Average pH value is 8.2. The electrical conductivity of soil was low 0.22 mS/cm. Organic carbon context of soils were low (0.54) as compared to coastal saline soil (4.8). The soils were having high CEC (58.4) as compared to coastal saline soil (43.37). Also having high content of $CaCO_3$ (6.3 per cent) as compared to coastal saline soil having (5.0).

The Magnesium, calcium, sodium and chlorides cations and anions increases the hardness of ground saline water which is limiting factor for good growth of fish. However, the ionic compositions of ground saline water various from location to location as per the geographical, climatic and hydrological conditions.

Scope for Aquaculture

The various state agril. Universities mainly Konkan Agril. University and W.B. Agril. University have developed low cost techniques to improve salt affected soils for productions of Agril crops, however, applicability is very limited. Further inland saline area is increasing every year. This water logged and salt affected area is lying unutilized for the agricultural crops. It is therefore very essential to concentrate efforts as culture of euryhaline fishes and prawn in low saline soils. The fisheries faculty of Konkan Agril. University has done lot of work in respect of suitability of culture of fishes / prawn in coastal saline soil locally known as "Khar Land ".

Coastal saline soils are mostly monocropped and rice is grown during Kharif. Inspite of heavy rainfall, occurrence of dry spells are common and maximum average dry spells of 13, 17 and 31 days were recorded in month of August, September and October respectively. To overcome this problem University has developed the surface dug out pond plays important role in integrated management of coastal saline soil *i.e.*, use of water for agriculture and aquaculture. The area required for this pond is 20 per cent of total agril land subjected to minimum water spread area of 0.20 ha. The depth of pond is 1 to 1.5 M below ground level and width of bundh is 3 M. The studies conducted on waters salinity are given in Table 8. The initial salinity of 14.5 ppt was reduced to 1.0 ppt after 5-6 years due to siltation. It has revealed that, this water can be utilized for culture freshwater fishes also.

Table 18.8: Salinity of 0.2 ha Pond (1.54 depth) in the Month of April at Panvel

Year	*Salinity*
1993-94	14.2
1994-95	9.2
1995-96	6.5
1996-97	6.2
1997-98	4.5
1998-99	2.5
1999-2000	1.0

Experimental Results of Aquaculture

Use of saline area for agriculture alone has certain limitation. It is therefore essential to concentrate efforts on aquaculture in these saline soils.

Rice-cum Fish Farming

In the simultaneous farming of rice and fish *Cyprinus carpio* yielded 500kg/ha under intensive culture 90 days and 149.1 kg/ha under extensive culture in 90 days (Shingare *et al.*, 1996).

Jitada (*Lates calcarifer*) Culture

It is fairly wide spread in the coastal area of the state. Under the combination of Jitada, Indian Major Carps and Tilapia, the production was 1475 kg/ha in seven months culture period (Shingare *et al.*, 1995)

Freshwater Prawn/ Shrimp Culture

With the identical technique of phased fertilization, the fresh water prawn *Macrobrachium rosenbergii* grew to an average weight of 60 g in 15 weeks while the tiger prawn *P.monodon* grew to an average weight of 43 g in 25 weeks (Shirgur, 1986). In nursery ponds at Pargaon post larvae of *M. rosenbergii* were reared by administering different combinations of ground oil cake, rice bran and fish meal, when maximum growth/survival was observed with 80 per cent oil cake + 10 per cent rice bran + 10 per cent fish meal (Shirgur *et al.*, 1990).

Conclusion

These results reveals that there is a potential for the above fish and prawn species in the ponds of saline soils.

References

Anonymous, 1990. Report of the Committee appointed to examine the present status of Khar Land Development. Government of Maharashtra.

Joshi, R.G. 1985. Characterization of the coastal salt affected soils of Maharashtra. Ph.D.thesis submitted to the Konkan Krishi Vidyapeeth, Dapoli.

Patil and Sonar 1994. Status of major and Micronutrients of Swell–Shrink Soils Of Maharashtra. *J.M. S. Agri. Univ.* 19 (2): 169-172.

Sahu, R.K. K.N. Chavan and S.B. Kadrekar 1981. Some hydrological observations of coastal saline soils of Konkan. 2nd Conference of "Problems and management of coastal saline soils" held at Bombay. Published by C.S.S. R.I. Canning Town (W.B.) pp 199-207.

Shingare, P.E. G.A. Shirgur and V.B Mehta 1995. On intensive culture trials of common carps in summer paddy plots in raigad district. Sem on rice in 21st century at Regional Agriculture Research Station, Karjat, March 1995.

Shingare, P.E., R.K. Singh, G.A. Shirgur and U.B.Mehta 1996. Effect of survival and growth of Indian Major Carps cultured with *L. calcarifer* and *O. mossumbicus* under mixed farming. Nat. Sem. on development of agriculture on west coast of India, Ratnagiri, Jan, 13-14, 1996.

Chapter 19

Effect of Water Probiotic on Water Quality and Growth of Catla Seed in Rain Fed Ponds

P.E. Shinagre, S.J. Meshram and S.S. Dhane

Introduction

The Indian Major Carps are commercially cultured for the production in India. In culture system, water quality plays an important role to maintain the health of fishes. As far as Konkan region is concerned, most of the village ponds are rainfed. The farmers of Khar Land Region are having small ponds of the size varying from 0.02 to 0.20 ha., which are called as surface dug out ponds. These ponds are preferably constructed to irrigate the paddy crop during dry spells.

The coastal soils are spread over in two climatic zones. The very high rainfall non-lateratic soils zone (VRN) comprises the Thane and northern part of Raigad District, where the soils derived from basaltic parental material, while very high rainfall lateritic soil zone (VRL) comes under southern part of Raigad district and whole district of Ratnagiri and Sindhudurg. The total rainfall is very high varies from 3000-4000 mm. However due to occurrence of dry spells in month of Aug, Sept and Oct., it is very difficult to protect the paddy crop.

Under such situation these surface dug out ponds plays an important role to irrigate the paddy crop during dry spell. Further, water during mansoon season remain as good as freshwater with zero ppt salinity.

The stored rainwater of these ponds is used for production of seed of Indian Major Carps during monsoon season, and it will be supplementary income to the farmers. Later on, these ponds can be used for growing the fishes to marketable size.

Being stored rain water and no facilities of other freshwater source, it is very important to maintain the water quality of these ponds. The input of supplementing feed and fertilizers also plays major role in producing good and healthy seed of Indian major carps. However, if proper care is not taken, this food may create unfavourable condition by producing harmful gases due to decomposing.

The production of these harmful gases can be controlled by providing aeration, water exchange and treating water with lime etc. However, it is labourious and time consuming. Further, there is no freshwater source for water exchange. Therefore, the water probiotic, containing *bacillus sp., Pseudomonas Sp. lactobacillus Sp. nitrosomonas sp.* nitrobacteria comprises stabilizer, nutrients and minerals etc. was used.

No such trials are conducted in the region, for production of seed of Indian Major Carp. Therefore the present study was carried out to observed the role of this probiotic on water quality parameter and growth of seed of Catla in rain fed nursery ponds. Since this type of probiotics are not used in freshwater fish, culture it's finding would be of great importance in nursery management practices.

Materials and Methods

The present trial was undertaken in rainfed nursery ponds having 0.02 ha water spread area. The seed of Catla with size range and average size 11 mm to 15 mm (12.6 mm) was stocked at the rate of 20 lakh/ha. The seed was well acclimatized before stocking in the nursery pond. The initial dose of probiotic was used at the time of stocking of seed and the second dose was applied after 15 days. The rate of application of probiotic was 25 kg/ha. The water quality parameters such as temperature, transparency, depth, pH, DO, CO_2, EC, Alkalinity, hardness, ammonia, nitrate and nitrite were measured weekly (APHA, 1989). The experiment was conducted for the period of 30 days. The growth was recorded at the end of experiment. The organic input such as groundnut oil cake and rice bran was used as feed. While Groundnut oil cake, Single super phosphate and urea was used as liquid phased fertilizer.

Results

The hydro biological parameters were recorded during the experimental period are given in (Table 19.1). The observations on body length, body weight and survival are given in (Table 19.2).

Table 19.1: Stocking Details of Catla Seed

Sl.No.	*Particulars*	*Experimental Pond*	*Control Pond*
1.	Water spread area	0.02 M^2	0.02 M^2
2.	Mean water depth	1.00 M	1.00 M
3.	Source of water	Rainfed	Rainfed
4.	Source of seed	Govt. seed farm, Khopli	Govt. seed farm, Khopoli
5.	Duration of experiment	30 days	30 days
6.	Initial avg. length(mm)	12.6 (11-15)±0.11	12.6 (11-15)±0.12
7.	Initial avg. weight (g)	0.085 (0.070-0.090)±0.18	0.85 (0.070-0.090)± 0.18
8.	Seed Rate per ha	20 lakh	20 lakh

All figures are shown as hectare basis.

Table 19.2: Range and Average of Various Water Quality Parameters

Sl.No.	*Water Parameter*	*Experimental Pond (Range)*	*Control Pond (Range)*
1.	Temperature (°C)	27.5-30.0 (28.5)	27.5-30.0 (28.5)
2.	pH	8.0-8.7 (8.5)	7.9–8.8 (8.2)
3.	Electrical conductivity (mmhos/cm)	0.32-0.33 (0.32)	0.32-0.33 (0.32)
4.	Dissolved oxygen (mg/l)	4.8–6.0 (5.4)	4.2–5.4 (4.9)
5.	Hardness (ppm)	290-340 (320)	305-360 (300)
6.	NH_4–N ppm	0.15–0.25	0.15–31
7.	NO_2–N	0.16–0.22	0.18–0.27
8.	NO_3–N	0.42–0.64	0.38–0.62
9.	Transparency	17–26.5	15–35
10.	Alkalinity	155–168	150–196

Table 19.3: Organic Input and Growth Details of Catla Seed

Sl.No.	*Particulars*	*Experimental Pond*	*Control Pond*
1.	Fertilizer		
	a) Lime	250 kg	250 kg
	b) SSP	75 kg	75 kg
	c) Urea	25 kg	25 kg
2.	Oil emulsion		
	a) Hyoxide–011	75 ml	75 ml
	b) Diesel	100 lit.	100 lit.
3.	Feed wed	GOC + RB	GOC + RB
4.	Water probiotic	25 kg	Nil
5.	Final avg.length (cm)	3.60 (3.0 to 3.6)±0.16	3.19 (2.9 to 3.2)±0.18
6.	Final avg. weight (g)	0.359*0.253-0.472)±0.18	0.256 (0.197–324)±0.19
7	Survival per cent	63.5	52.0

All figures are shown as hectare basis.

Abbreviation: ±: Standard error of mean; *: ($P<0.05$)

The temperature of both the pond was same while higher values of pH were recorded in control pond as compared to experimental pond. Range of dissolved oxygen is higher in experimental pond with an average of (5.4 ppm) as compared to control pond (4.9 ppm) Ammonia and $N0_{2-}N$ concentration were found to be more in control pond as compared to experimental pond.

Thirty days of feeding showed significant growth in body weight ($P<0.001$) in experimental pond as compared to control pond. Survival percentage was higher (63.5 per cent) in experimental pond as compared to control pond (50.2 per cent).

Discussion

Results of the present investigation indicated that the growth of Catla fry was far better in waters treated with probiotic. This also indicates that the bacterial strains and substrains present in probiotic played an important role in the enhancement of weight of fry of Catla. Kumar (1994), Sharma and Kumar (1998) reported significant impact of microbial populations on the growth of fish. Autotrophic and hecterotrophic bacteria plays important role in the growth of fishes (Schroeder, 1978). Fingerling of *C. mrigala* showed good growth fed with fishmin (Sharma and Kishan, 1997). Sharma and Bhukhar, (2000) also observed good growth of *C. carpio* fingerling with Aquazyn TM-1000 at the rate of 1.25 mg/l.

Conclusion

Thus, from the foregoing discussion it could be concluded that the water probiotic plays an important role in the growth of *Catla* fry by improving and maintaining water quality of rainfed ponds.

References

APHA, 1989. Standard Methods for the Examination of water and Waste water. Am. Public Health Associate Washington, DC 17th edn., 1452 pp.

Kumar, A., 1994. Evaluation of Fishmin in relation to microbial and zooplanteton population and fish growth. M.Sc. Thesis, Rajasthan Agricultural University, Bikaner, 90.

Schroeder, G.L., 1978. Autotrophic and heterotrophic production of microorganisms in intensely fertilized fish ponds and related fish yields. *Aquaculture*, 14: 303-325.

Sharma, O.P. and R. Kishan, 1997. Efficacy off fishmin (a multimicronutrient fertilizer) in the growth of fingerlings of *Cirrhinus mrigala* (Ham.). *Ecol. Env. and Cons.*, 3 (3/4): 189-192.

Sharma, O.P. and A.Kumar, 1998. Growth of fingerlings of *Cirrtinus mrigala* (Ham.) in waters fertilized with fishmin. *Fishing Chimes*, 18 (5): 15-16.

Sharma, O.P. and S. Bhukhar, 2000. Effect of Aquazyn TM-1000, a probiotic on the Water quality and growth of *Cyprinus carpio* var *communis* (L.) *Ind. J. Fish*, 47 (3): 209-213.

Chapter 20
Biology of Anostracan Zooplankton

S.P. Chavan and M.S. Kadam

Introduction

Zooplankton is the secondary biomass in an aquatic ecosystem, zooplankton is one of the important food resource for the primary consumer level fishes and other planktophagous animals in an aquatic ecosystem. Along with phytoplanktons, the zooplankton desides the productivity of a water body. The phytoplanktons includes microscopic, unicellular, microfilamentous or microcolonial algal species. The phytoplankton species with the help of their chlorophil pigment, in presence of carbondioxide trap the solar energy radiations to produce the chemical form of energy *i.e.*, glucose molecules, this needs water and other mineral nutrients. Hence, the phytoplankton are the food factories in an aquatic ecosystem.

Zooplankton are free floating, suspended variety of animals, flow or drift with water current in the running water or in the lotic ecosystems, but in the stagnant or lentic ecosystem *e.g.*, pond, lake, reservoir, the zooplankton show the upwelling or up and down vertical movements inresponse to the temperature and solar radiations called as vertical movement of zooplankton.

Zooplankton includes mainly the members from various classes of phylum–Arthropoda but the protozoans and adult, free swiming stages of some helminths, larvae of molluscs, planktonic eggs of many invertebrates also came in the list of zooplankton. The quality and quantity of the zooplankton in any freshwater body is desisive especially for its fisheries development. By the study of physico-chemical factors of a freshwater body the planktonic content could be corrected to the optimum requirement for the flourishment of fisheries in some extent.

Anostracan Zooplanktology

Introduction

Anostrace (Ralf w. Dexter, 1943, 1951) is one of the important order of

Class–Crustacea of phylum–Arthropoda

Sub class–Branchiopoda

Anostraca includes mostly freshwater zooplankton species, distributed in six (06) families mainly. These are

Family–Polyartimidae

Family–Streptocephalidae

Family–Branchinectidae

Family–Artemiidae

Family–Chirocephalidae

Family–Thamnocephalidae

Distribution of Anostraca

Fairy shrimps are sporadic in occourance or randomely distributed, one freshwater pool may have an abundant population but another nearby freshwater body may not have any shrimp. In one region these may be many while in another region there will not be any occourance. There is no any specific pattern of distribution. It was also observed that, their occourance at a place may fluctuate year to year. In most of the parts of the world, fairy shrimps occour in freshwater ponds. Fairy shrimps can withstand predation from insects and amphibians but are soon eradicated by the fishes but recently in Maharashtra state of India in few reservoirs of Parbhani District, specially in the Masoli reservoir (Medium size, 360 ha) remarkabely the co-existance of fairy shrimp and fishes was observed during summer season and it was another remarkable feature of occourance that, the *Branchinecta* genus was found in such water body in the temperate zone of the world, as compared to earlier species of the Genus–*Branchinecta* recorded from North America, Arctic and Alpine zones of the world. Rainfed water–bodies and melted snow ponds are one of the common habitats of fairy shrimps. The fairy shrimps swim gracefully on their bucks, appendages are always faced towards the source of light.

Mature specimans frequently rests o their back at the bottom sediments of water body.

Colouration

Body colour of fairy shrimps extremely variable, differs from place to place, from one life stage to the next. Kind of food ingested by them also desides their colour sometimes.

Size

North American fairy shrimps range in the size form 5-6 mm to 60-100 mm (*Eubranchipus floridanus* and *Branchinecta gigas* respectively). Most of the species are in the average size about 20-25 mm in total length.

Reproduction

Sexes are separate and internal fertilization is ordinarily required for the development. Males are less abundant than females. Gonads of both the sexes are paired, tubular organs. The fertilized eggs develop in the egg sac and may be released into the water body or remain in the eggsac of female at its death. The eggs are distributed by the wind and through the mud carrying animals visiting the pond *e.g.* aquatic birds, cows, buffaloes, antlers, deer, newts, frogs etc.

After hatching from the cysts or fertilized eggs the nayplius larva oozes out into the water, in which many moults take place to develop into the adult, hatching of the cysts needs the drying or decication period as a stimulus of development.

Methodology

By using 150 mm conical plankton net, at the morning time, from the coastal water of a pond or reservoir the adults of fairy shrimps can be collected by netting at 1–1.5 meter depth very successfully. Collected specimens along with other zooplanktons cannot be isolated, hence whole zooplankton mass is preserved, fixed in 4 per cent formalin or 1: 1 mixture of methyl alcohol and formalin, diluted to 4 per cent concentration. Then dehydrated and stained by alcoholic eosin and other stains and micro-preparations are prepared; to study the further taxanomic details under common compound microscope or computer aided phase contrast microscope.

To study the scheme of classification, mainly the reproductive organ of male, thorasic appendages and head appendages are observed under compound microscope by using reflecting lenses sometimes third hand was sever from the body to observe their detail features.

General Features of Anostraca

1. The Anostracans are also called as fairy shrimps, these are the most primitive crustaceans of the recent fauna.
2. They are elongated, somewhat cylindrical, delicate animals without carapace.
3. The head is large and prominent, bears sensery organs, clasping organs in male and various mouthparts; compound eyes are large and stalked.
4. In the middle of the head, there is a single sessile eyespot called as ocellus.
5. The first antennae are slender, small and inconspicuous, often they are unsegmented.
6. The second antennae are large and swollen, these are enormousely developed in males as compared to the females, sometimes there are antennal appendages in the form of processes, tubercles and spines.
7. Each species has developed unique characteristics of the second antennae, considered as basic feature in the classification.
8. Sometimes there is a frontal appendage, attached to the head between the second antennae which is presumed as the claspers; this is an important base for the classification of *Anostracans.*
9. The mouthparts consists of overlapping labrum, pair of mandibles and two pairs of maxillae. The buccal cavity oesophagous is vertical, globular stomach and digestive glands are located within the head.
10. A long streight intestine runs to the extreme posterior end of the animal body.
11. Thorax contain 11 or 17 segments, each with a pair of follacious appendages, these serve for locomotion, feeding and respiration. The append ages which are biramous, lobed and with setae are much alike, last pair of appendage is not fairely developed. Laterally there are one or two pre podites fused to varying degrees.
12. On the base of appendage there is epipodite, which never has spines; distally there are exopodites and endopodites, along the medial margin, there are fine endies bearing setae as well as filaments.
13. Following the limb bearing thorasic segments, sometimes called as pre-geintal region or trunk, there are two partly fused thorasic segments which contains reproductive organs. In

the male there are two penis the structure and position of penis are taxonomically important. In the female, there is well developed ovisac, in which the eggs develops posterior to the genital segments there are seven abdominal segments also called as post genital segments, the last segment called as telson. It carries two cercopods; Sometimes on (In Indian fairy shrimp, Chavan *et al.*), these are useually elongated and armed with spines or long filaments.

Key to the Species

(By the Drs. Folk Linder, Walter G. Moore, J.G. Macklin, N.T. Mattox)–F.W. Biology, H.B. Ward, G. Chandler Whipple).

Here, only key to the species of the Genus–*Branchinecta* are given; from the Family–Branchinectidae (Verrill, 1869).

Family–Branchinectidae–penes (pair), widely separated from each other, on the later surface of the body Genus–*Branchinecta* (Verrill, 1869). This is only one genus, of the family Branchinectidae; proximal segment of male clasping antenna serrate on inner margin.

Species of the genus *Branchinecta*–Uptill now 20 species of the genus *Branchinecta* were reported by different workers on Anostracan taxonomy, last species. *B. parmahaindiensis* was reported from India in 2005. All reported species were errected as new species by comparing their body features, as explained under the heading–General body features in the present topic.

The species of the Genus–*Branchinecta* with their common name are given below:

1. *Branchinecta paludosa* (O.F. Muller, 1788) circumpolar fairy shrimp.
2. *Branchinecta coloradensis* (Packard, 1874) Colorado fairy shrimp.
3. *Branchinecta lindahli* (Packard, 1883) Versatile fairy shrimp.
4. *Branchinecta packardi* (Pearse, 1912) rock pool fairy shrimp.
5. *Branchinecta gigas* (Lynch, 1937) Giant Fairy shrimp
6. *Branchinecta shantzi* (Makin, 1952) Newcolorado fairy shrimp.
7. *Branchinecta Mackini* (Dexter, 1956) Alkali fairy shrimp.
8. *Branchinecta cornigera* (Linch, 1958) Horned fairy shrimp.
9. *Branchinecta compestris* (Linch, 1960) packed pouch fairy shrimp.
10. *Branchinecta dissimilis* (Linch, 1972) Dissimilar fairy shrimp.
11. *Branchinecta potassa* (Belk,1979) Potassium loving fairy shrimp.
12. *Branchinecta conservatio* (Belk, Eriksonn, 1990) Conservency fairy shrimp.
13. *Branchinecta Lynchi* (Belk, Erikso, 1990) vernal pool fairy shrimp.
14. *Branchinecta longiantenna* (Belk, Erikson, 1990) long horn fairy shrimp.
15. *Branchinecta sandiagonesis* (Fugate, 1993) sandiago fairy shrimp.
16. *Branchinecta kaibabensis* (Belk, Fugate, 2000) Kaibab fairy shrimp.
17. *Branchinecta massovallensis* (Belk, Fugate, 2000). Midvalley fairy shrimp.
18. *Branchinecta readingi* (Belk, 2000) eastern alkalifairy shrimp.
19. *Branchinecta hiberna* (Rogers, fugate, 2001). Winter fairy shrimp.

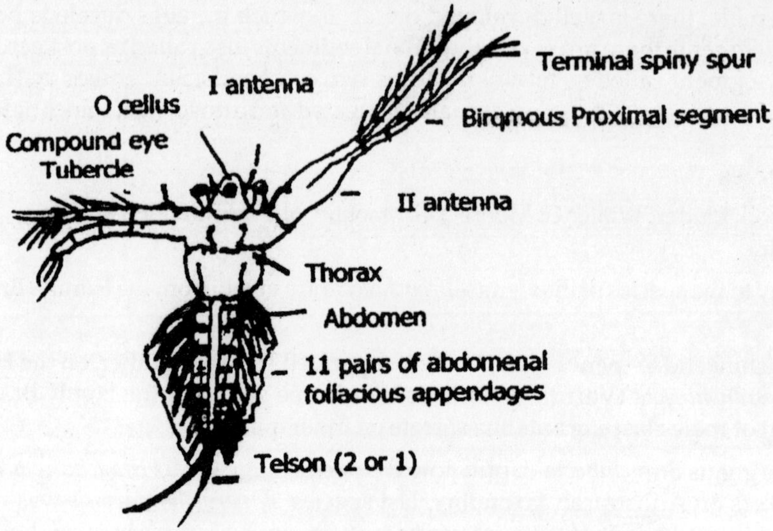

Figure 20.1: *Branchinecta* sp. (External Features)

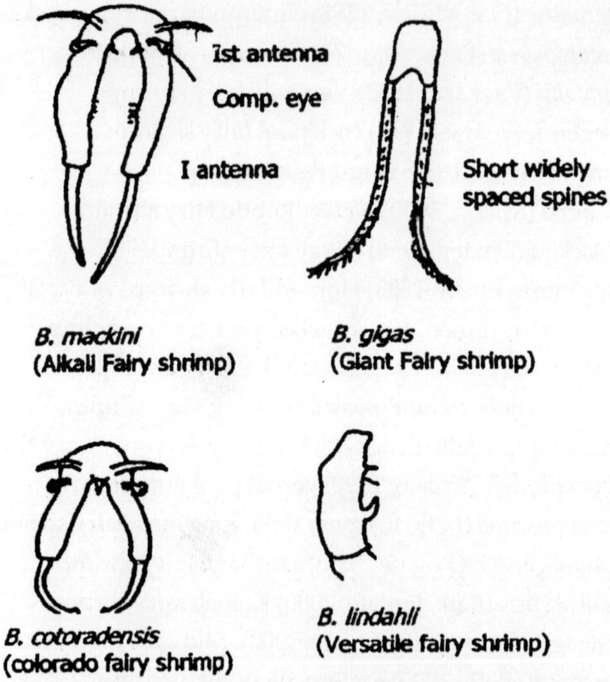

Figure 20.2: Some Species of *Branchinecta* with their 2nd Antennary Variations

Importance of the Study of Fairy Shrimps

Uptil now, there are many experimental trials were made by many workers on the collection of small, granular cysts of the brine shrimp–Artemia sp. and its culture in the laboratories; also it is now being available on commercial scale, because it was found that, the biochemical estimate of Artemia is very good, nutritions, specially proteinecious for the fishes under semi–itensive and intensive culture practice, the larval stages of the artemia are excellent food for some aquarium fish species.

There is no any attentive work has been done on the taxonomy, biochemistry and culture of freshwater fairy shrimps. It is necessary for the improvement of reservoir fisheries and pond culture fishery. The detail study of fairy shrimps could be a mile stone in the nutrition science of fish species like cyprinus and catla.

References

Bohonak, A.J., Whiteman, H.H. 1999. Dispersal of fairy shrimp *Branchinecta coloradensis*: Effects of hydroperiods and salamanders. *Amer. Soc. of Lim. Oceanogr*, 44 (3): 1999; 487-493.

Gallagher, Mozlev, A. 1932. Biological study of temporary pond in western Canada, American. *Nature*, 66. 235-249.

Gallayher, S.P. 1996. Seasonal occourance and habitat characteristics of some vernal pool Branchiopoda in Northen california. *J. of Crustac. Biol.* 16: 323-329.

Timms, B.V. 2001. Two new species of fairy shrimp (Anostraca) from Parro, Inland, Australia. Records of the Aust. Museum (2001), Vol. 53: 247-254, ISSN 0067-1975.

Ward, H.B., G. Chandler Whipple. 1959. Freshwater Biology (2nd Edition).–New York, London, J.W. and Sonss, New York.

Index